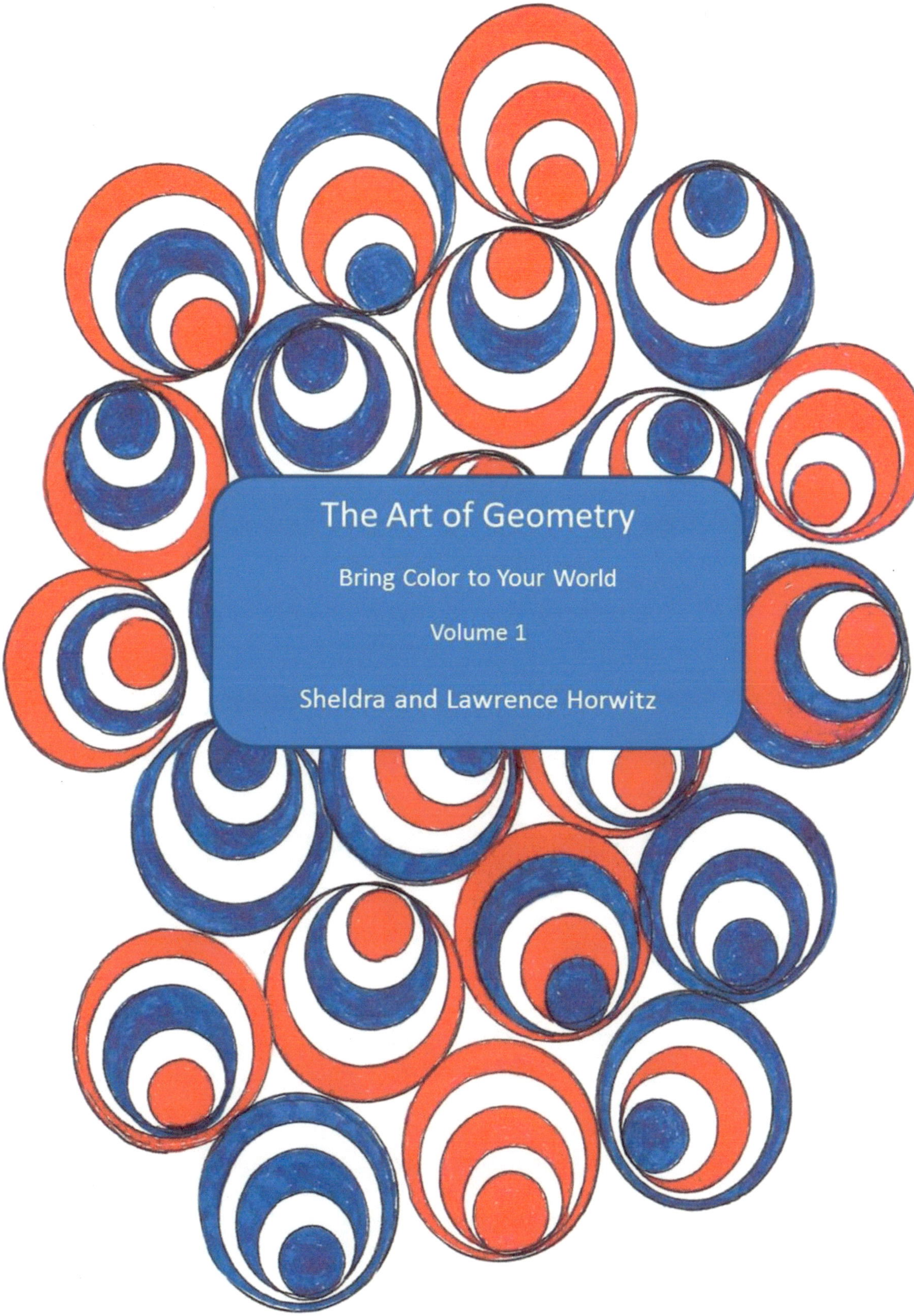

The Art of Geometry

Bring Color to Your World

Volume 1

Sheldra and Lawrence Horwitz

Penway Publications
Printed by CreateSpace, an Amazon.com company

ISBN: 0997159806
ISBN-13: 978-0997159806

shellyandlarry.com

Dedication

To Our Grandchildren

Lisa, Michael, Laura, Marcello

The Art of Geometry

WELCOME to the happy world of geometric shapes.

THIS IS a different kind of coloring book. Instead of filling in the shapes with color, we encourage you to study the colored examples provided throughout the book and then study full-sized blank versions that follow. Do you see the similarities? Sometimes it's not easy to see how they are related. Some have new lines or curves added. Some have only colored a portion of the picture to emphasize interesting shapes that have been formed by the lines.

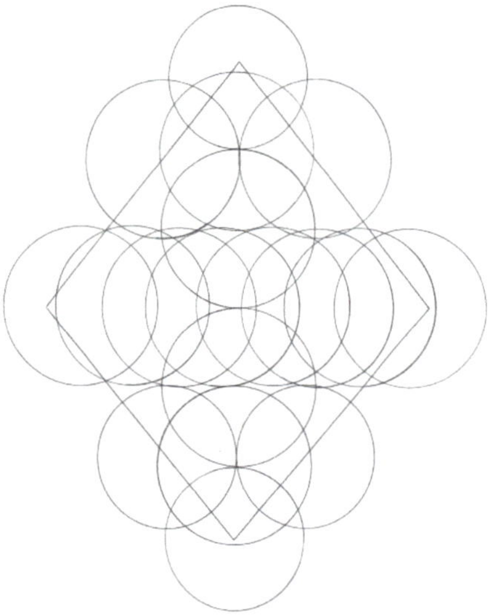

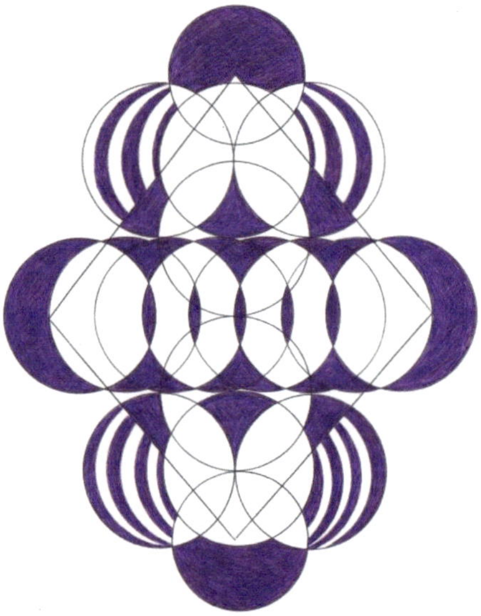

WE HAVE given you an idea of how the picture may be colored. However, imagination and creativity will allow you to enhance the figures to your liking. You may wish to use a ruler, compass, protractor and polygon stencils, French curves. You can also use a penny, a plate or any household objects with interesting lines and curves.

YOU WILL now be known as a "colorist". So take out your colored pens and pencils and let us begin.

Shelly and Larry

shellyandlarry.com

Example:

My husband creates the original design. This was made using a compass.

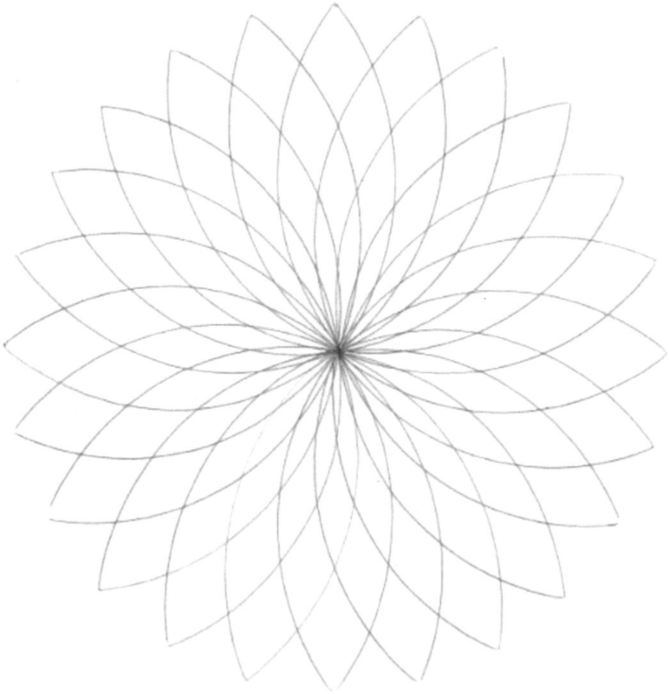

Take time to study it for a while.

Notice the small star in the middle.

Do you see all the shapes that are formed by overlapping curved lines?

Notice the large flower petals? Do some of the petals stand out more than others?

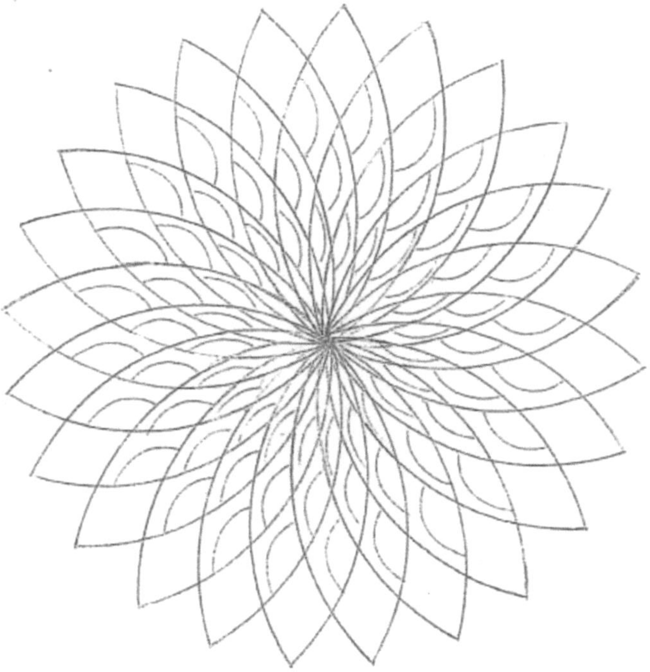

Maybe you will color just a few of the petals.

I added curves using a penny. Look around your house for interesting forms.

Notice the new shapes you create as a result.

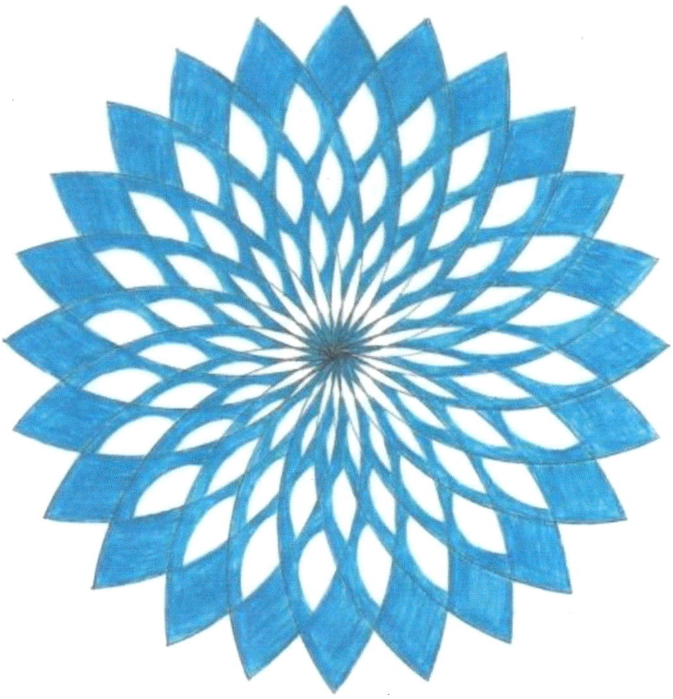

Look at the design again and choose what section to color and what sections to leave uncolored.

The areas you color will stand out.

This blue one has a soft, lacy snowflake look.

A darker color would show more contrast and highlight the areas left uncolored.

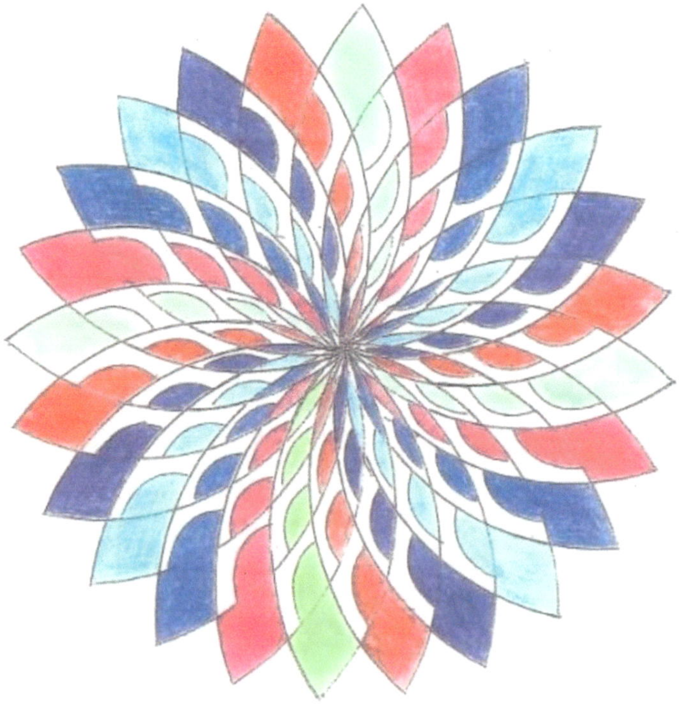

This is the same picture but has different areas colored giving it a completely different look.

Feel free to make copies of the blank pictures so you can try it different ways.

This coloring book gives you opportunities to experiment.

Most of the blank pictures have a duplicate on the reverse side – sometimes they are identical and sometimes the duplicate leaves even more opportunity for you to come up with your own unique design

Let’s get started . . .

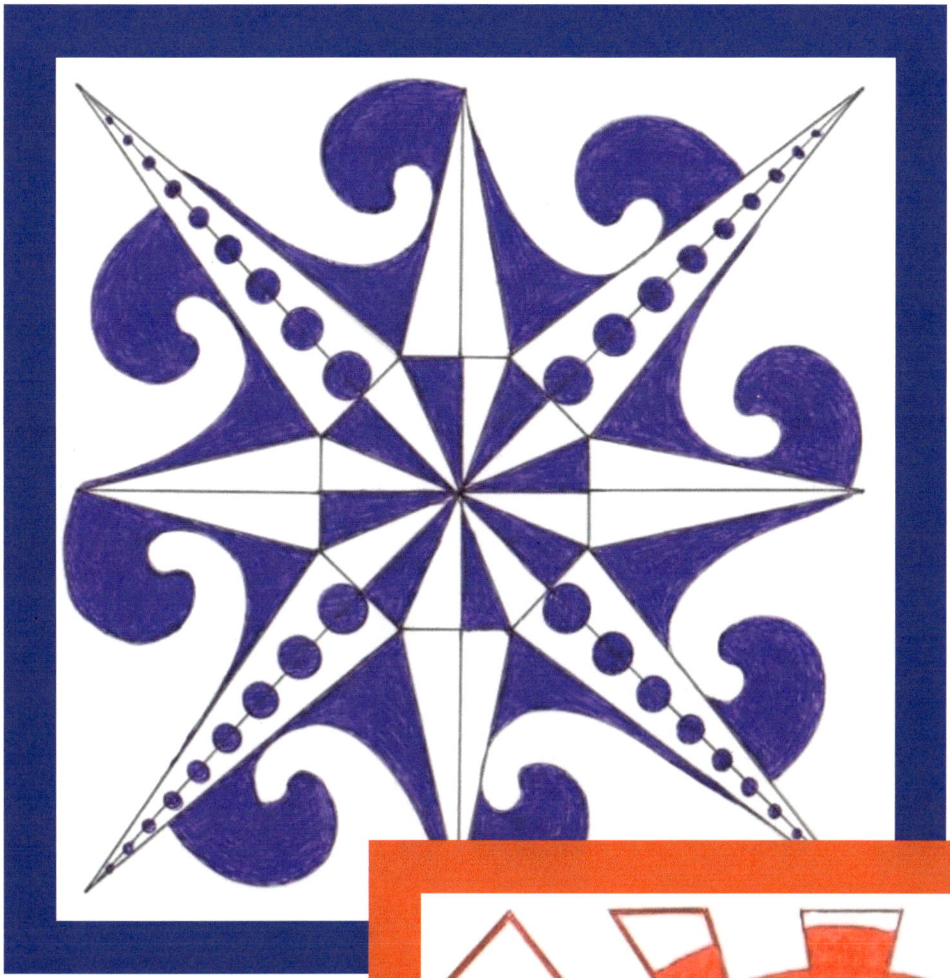

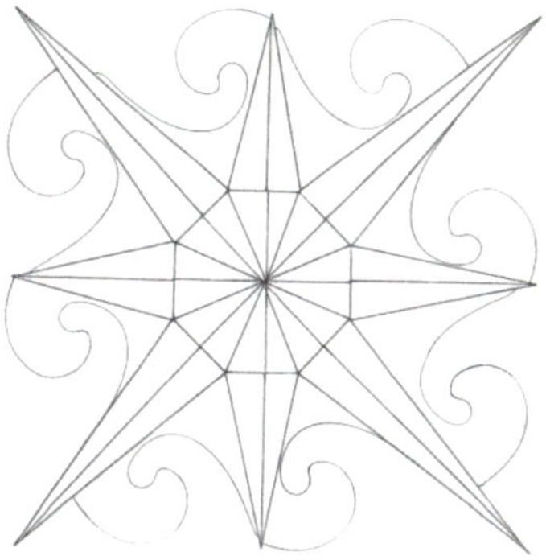

Royalty

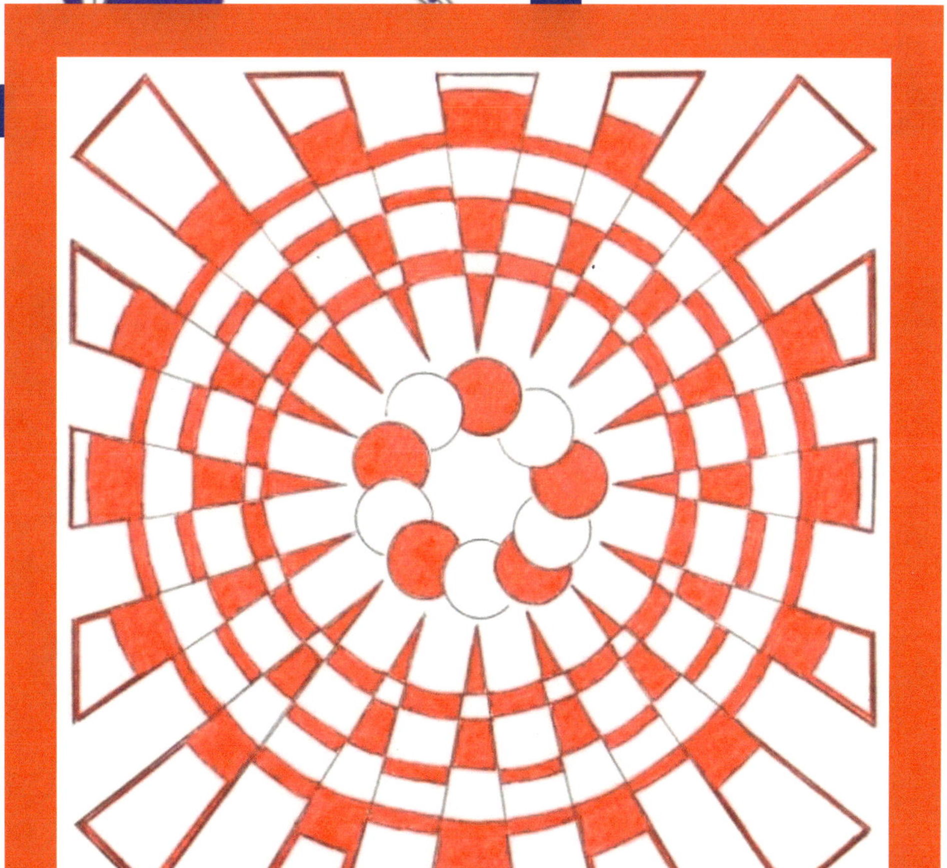

An Exciting Life: Enjoy

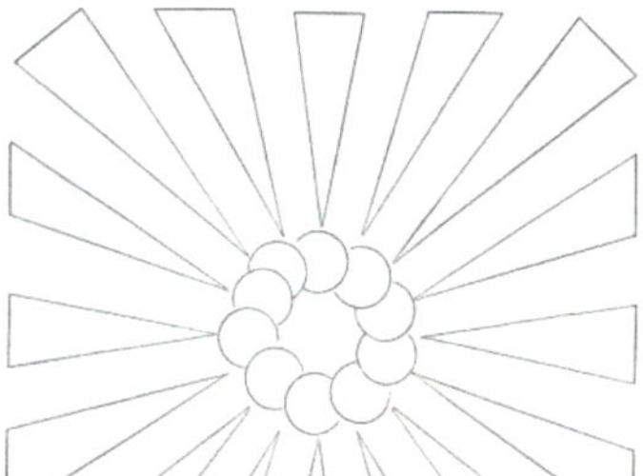

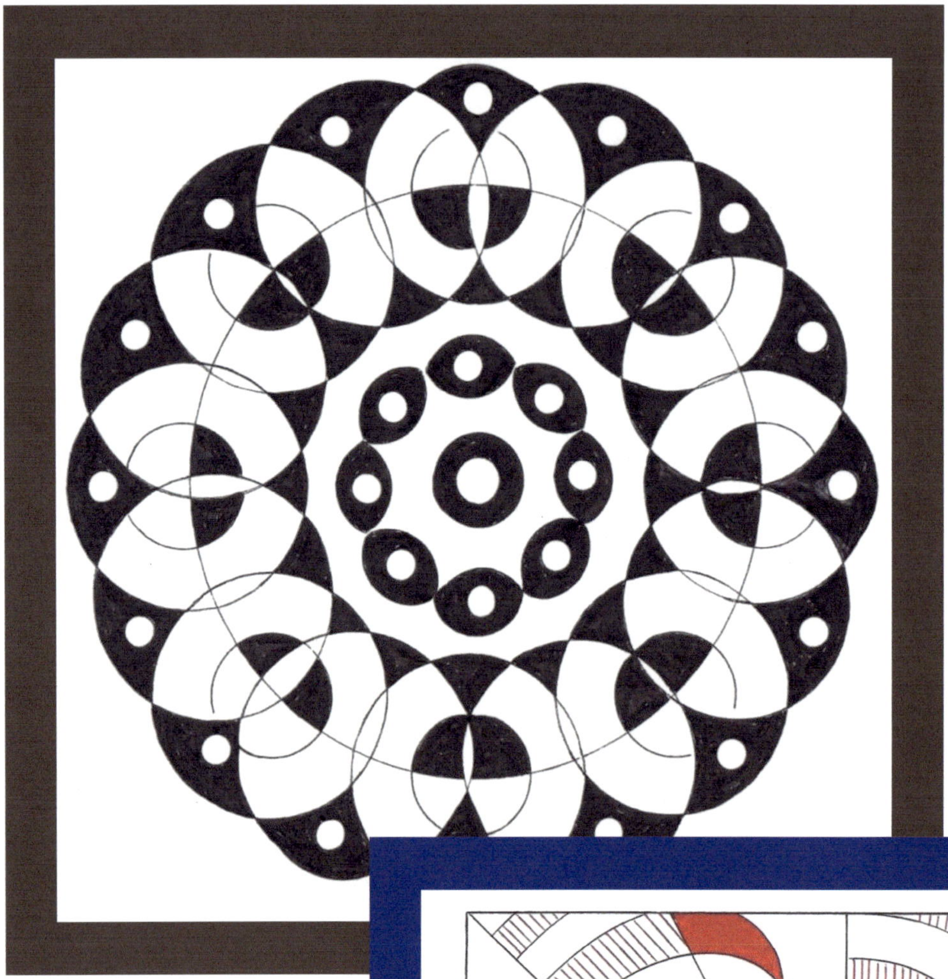

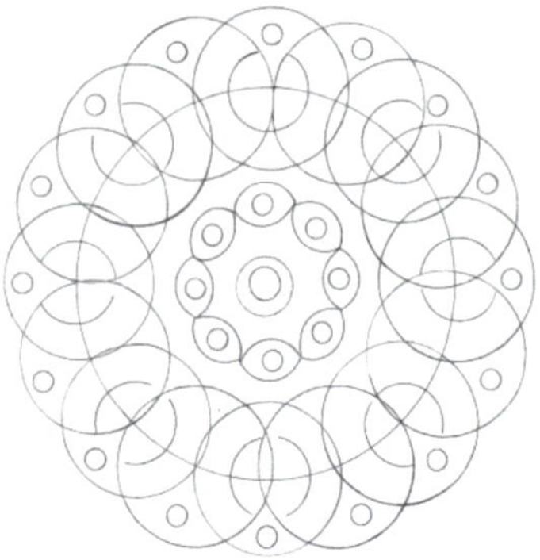

Midnight

Very American

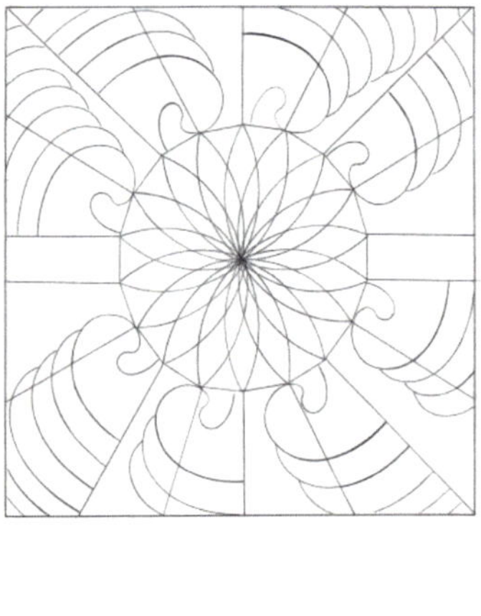

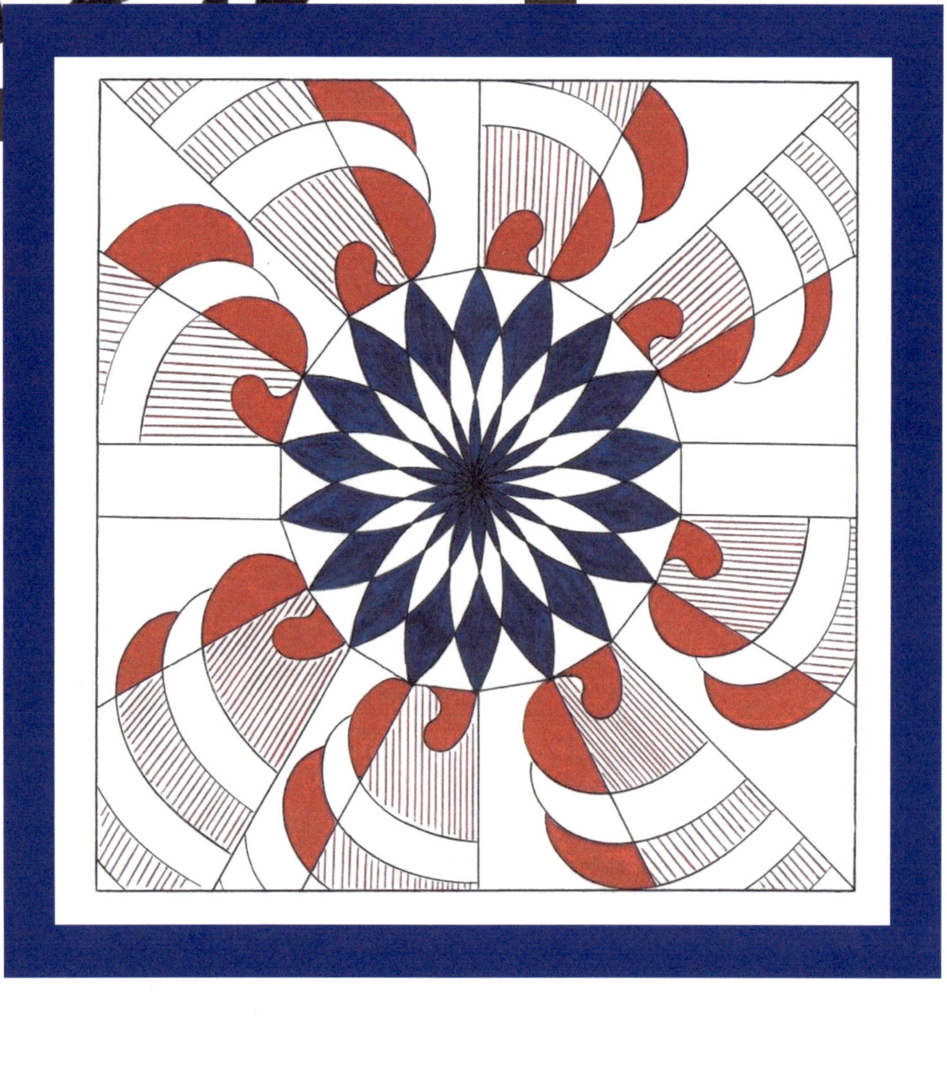

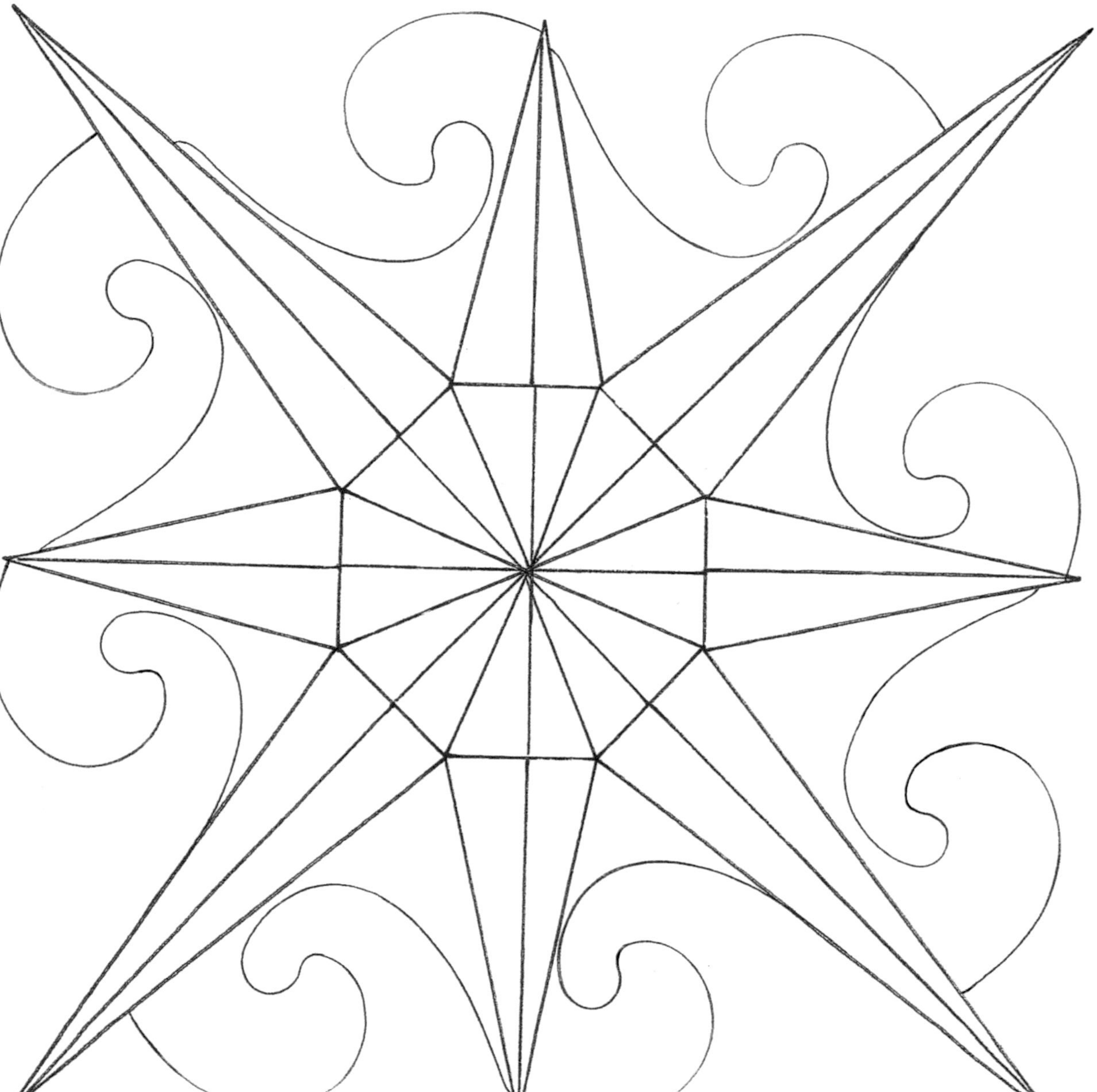

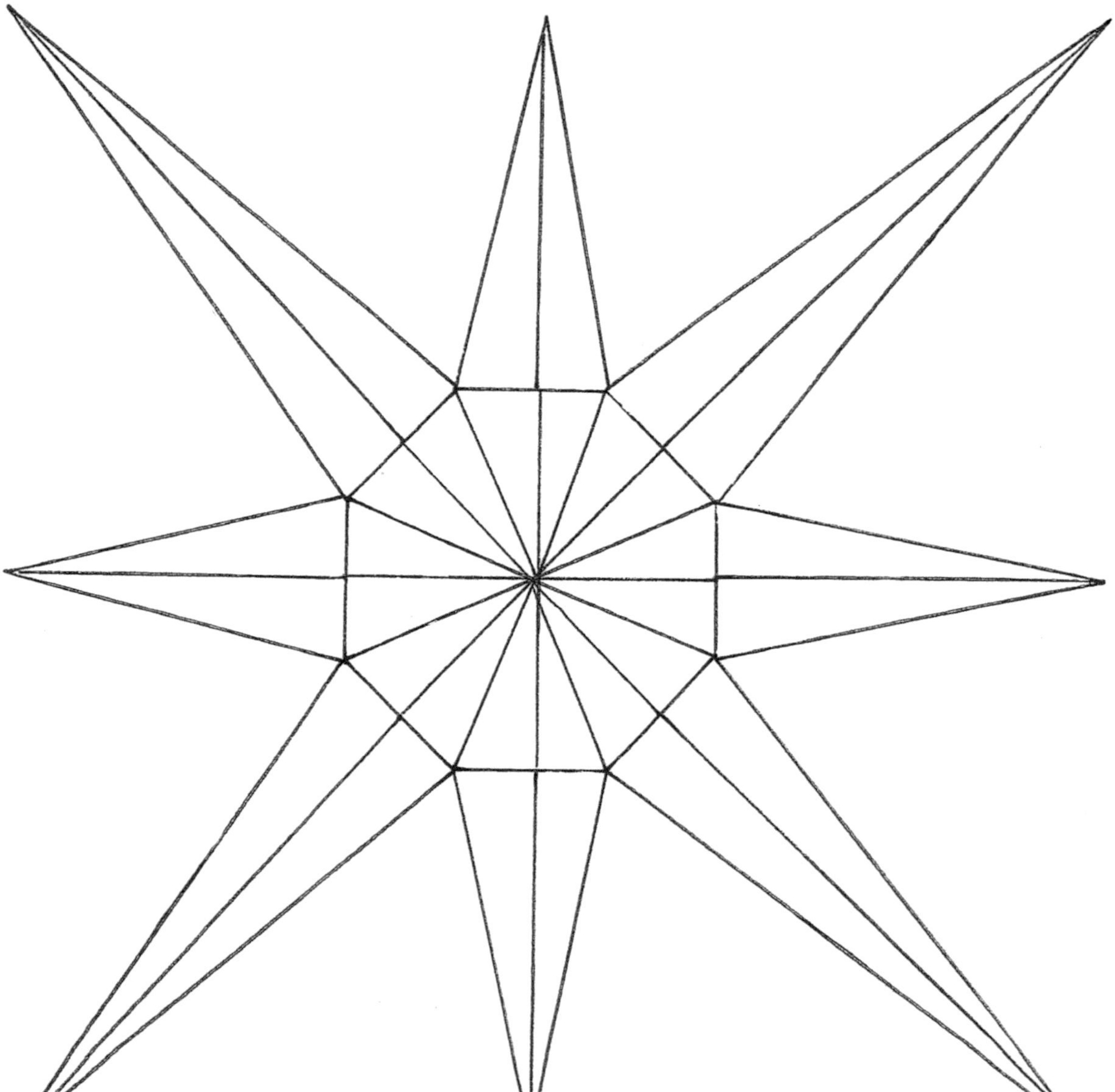

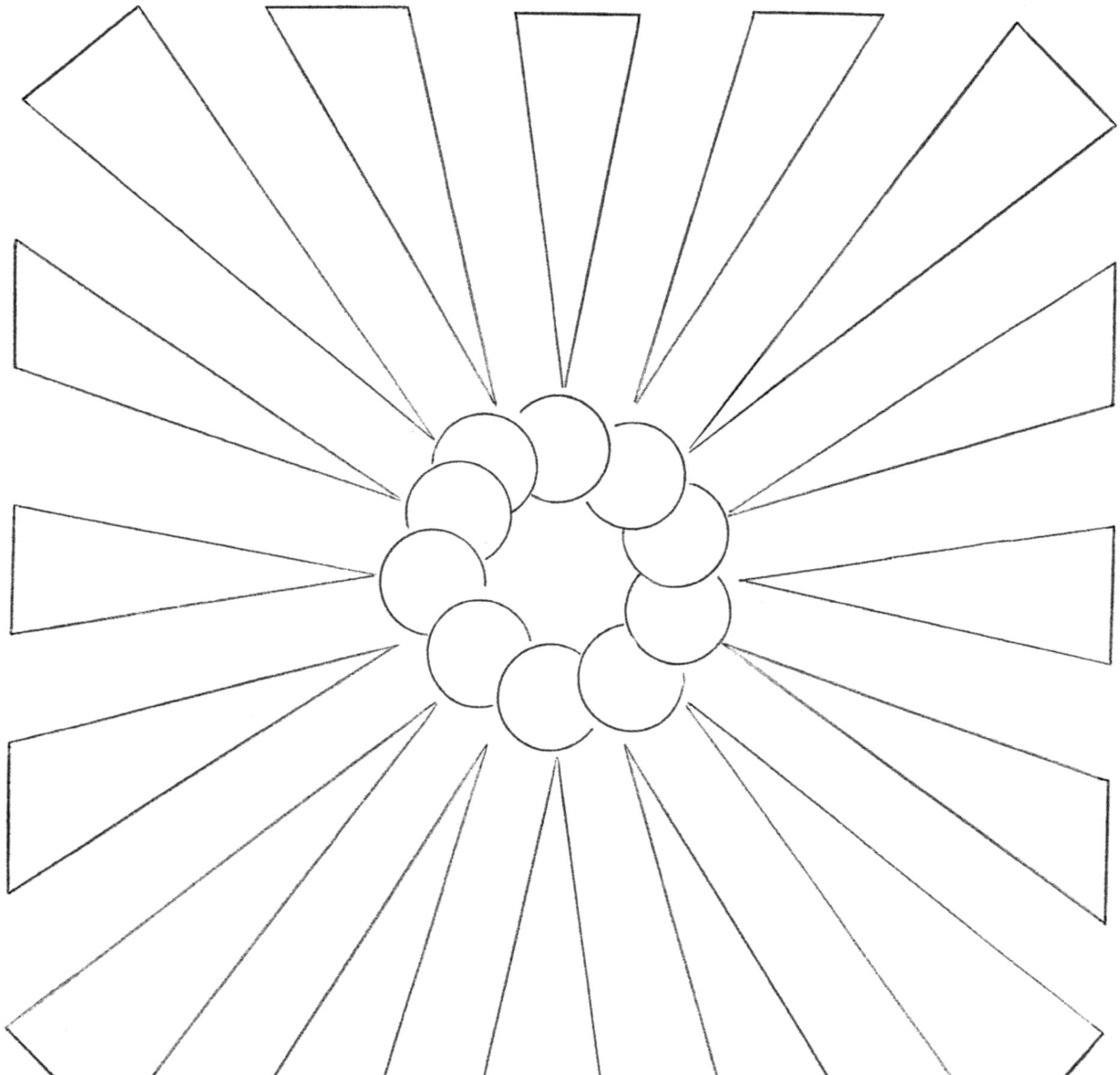

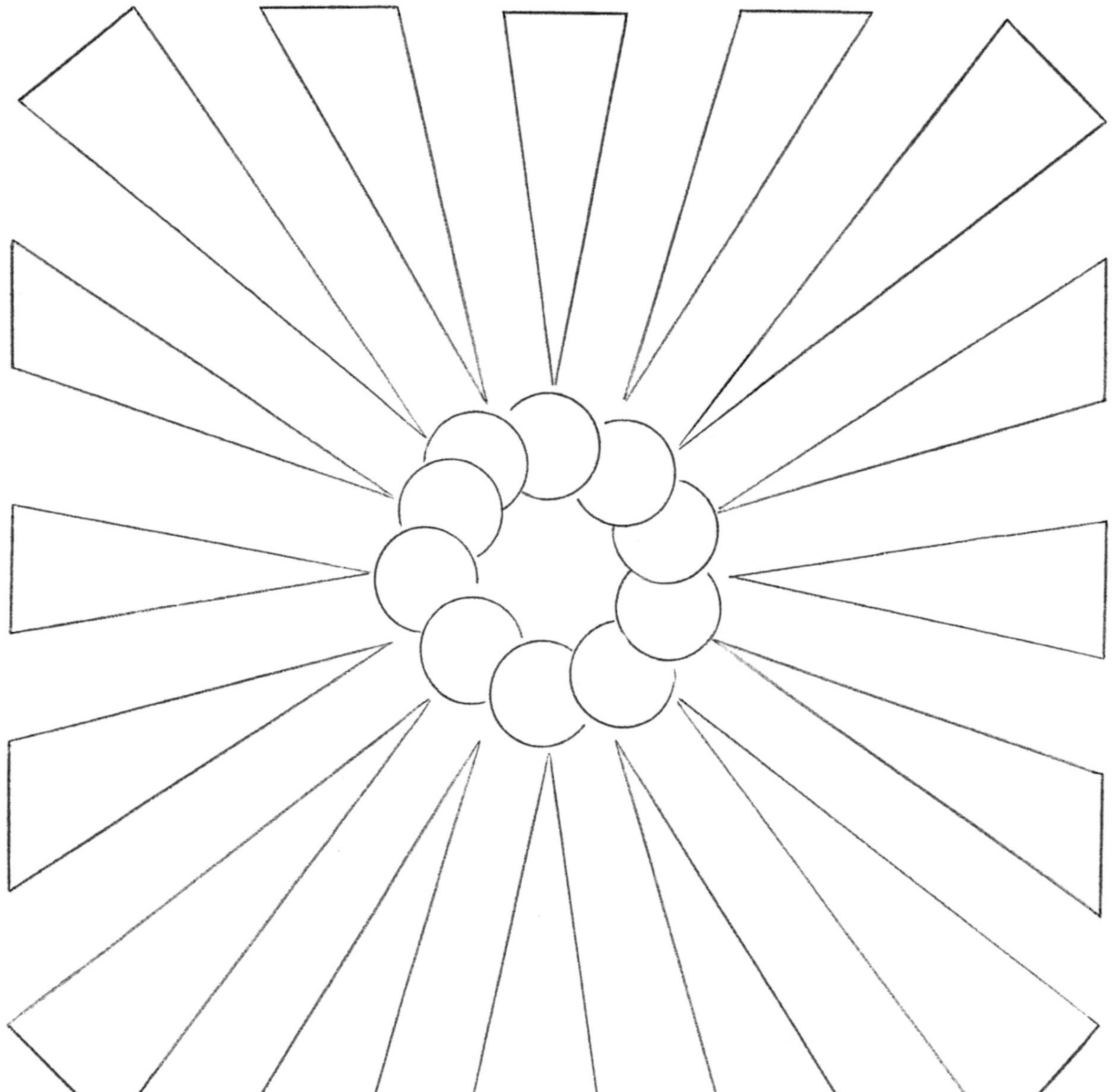

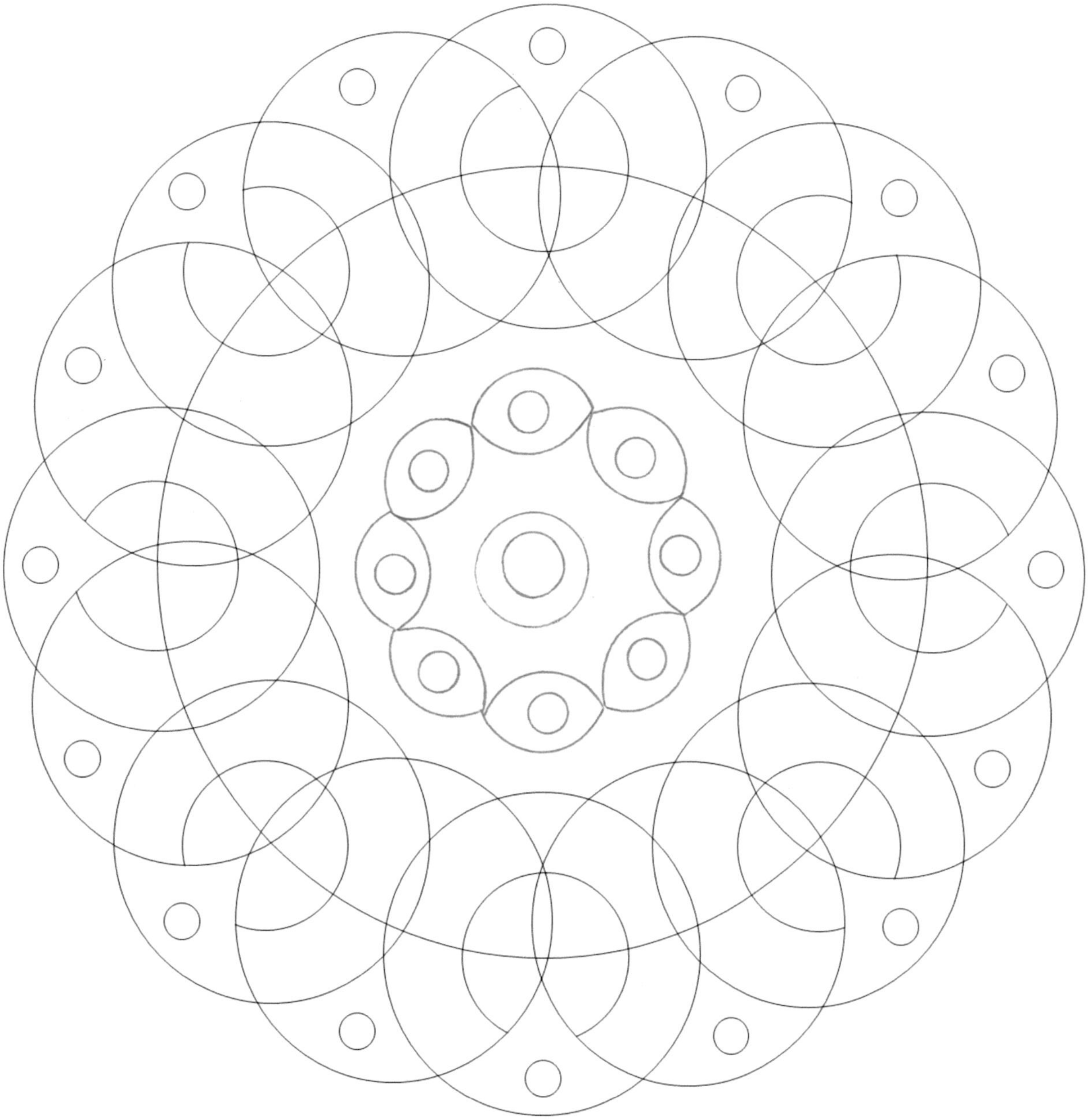

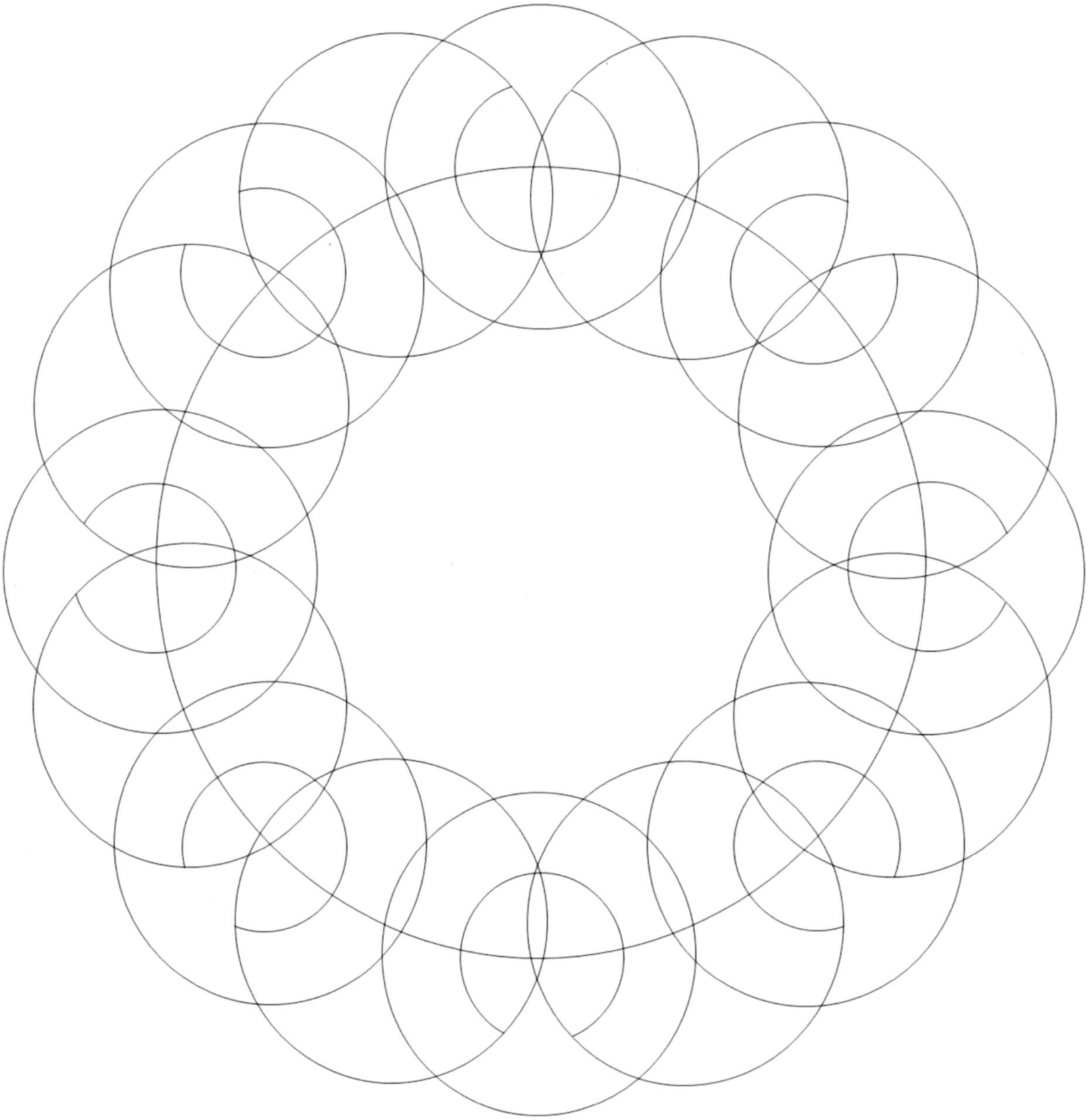

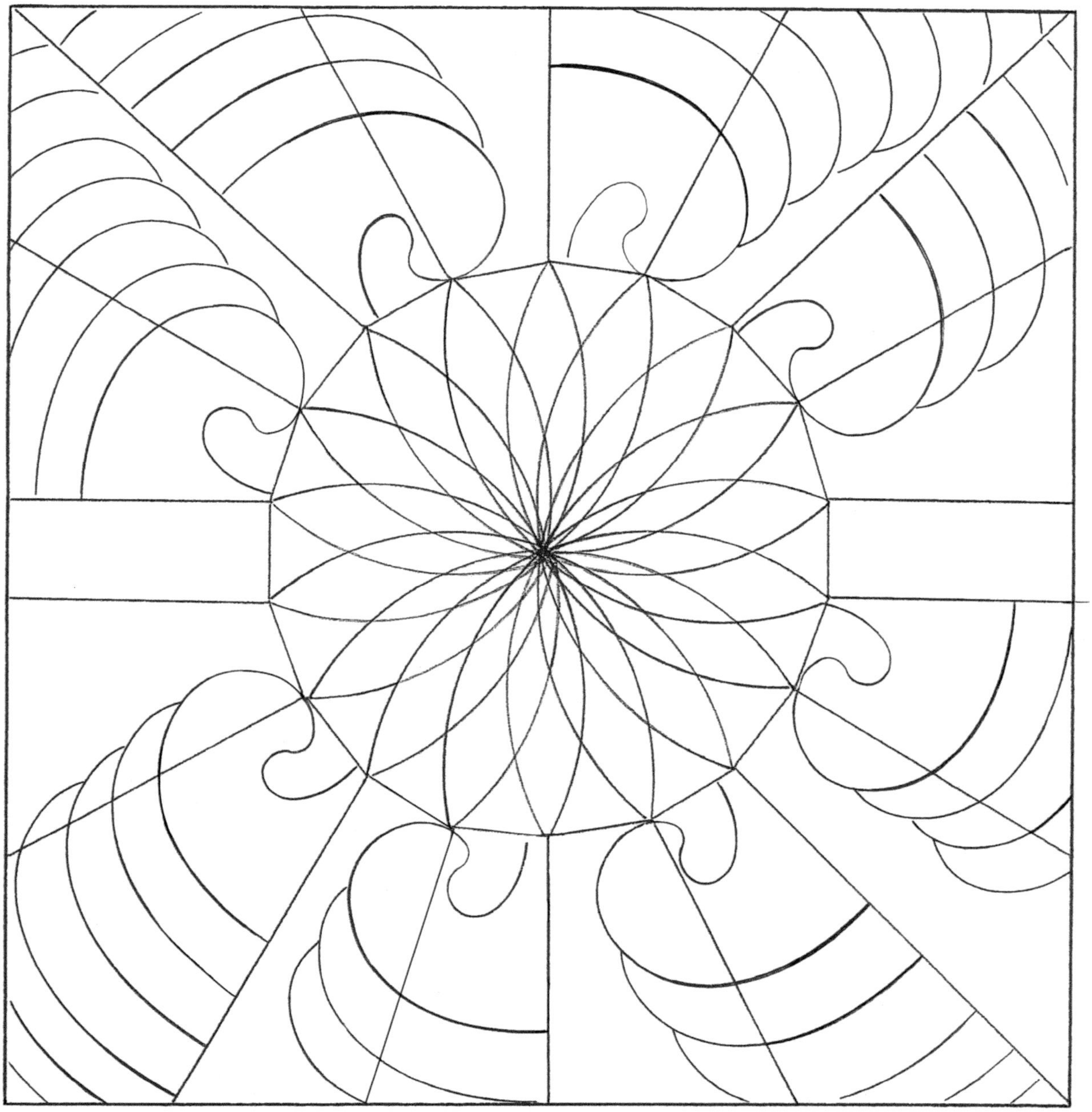

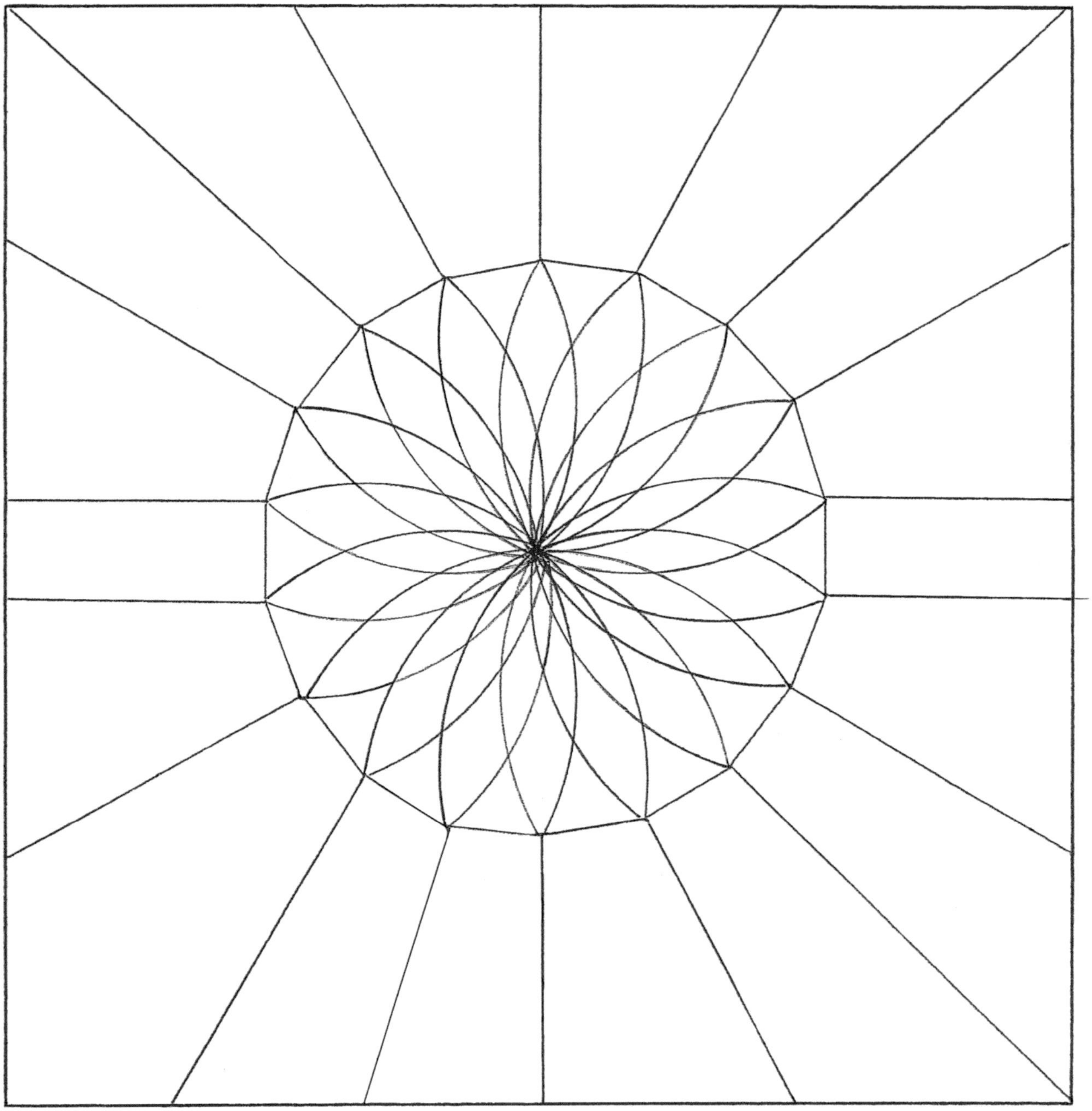

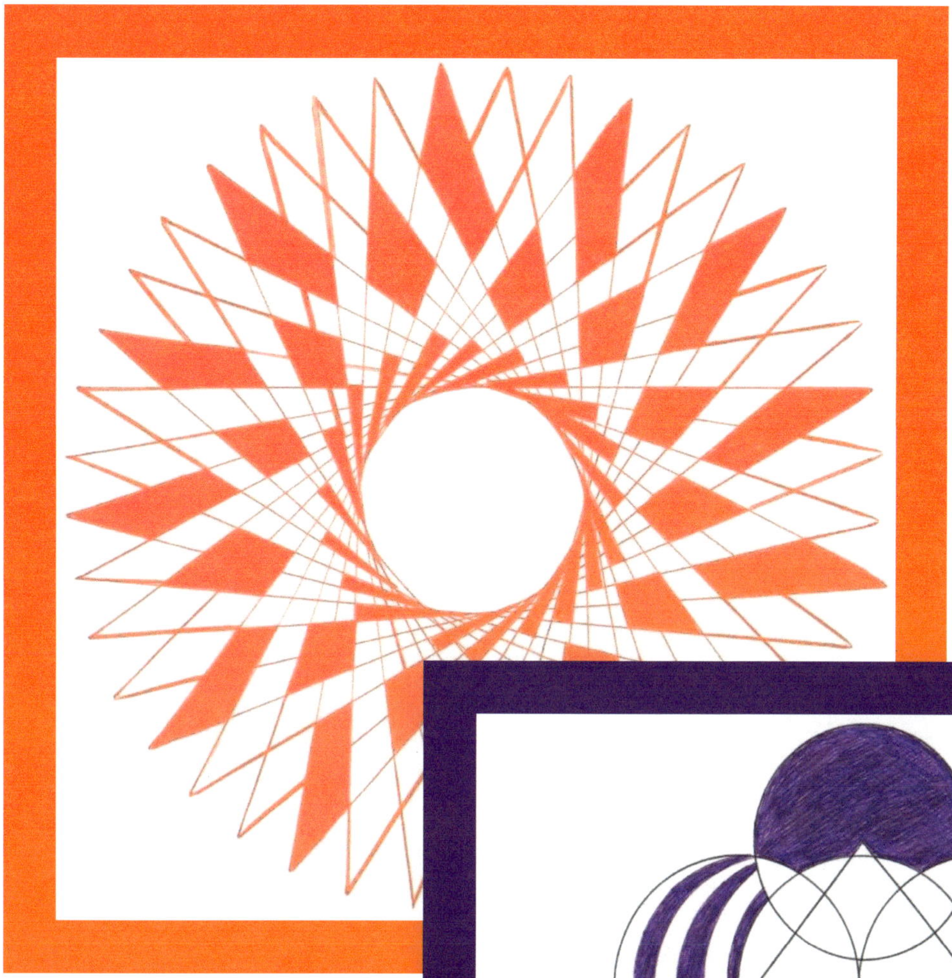

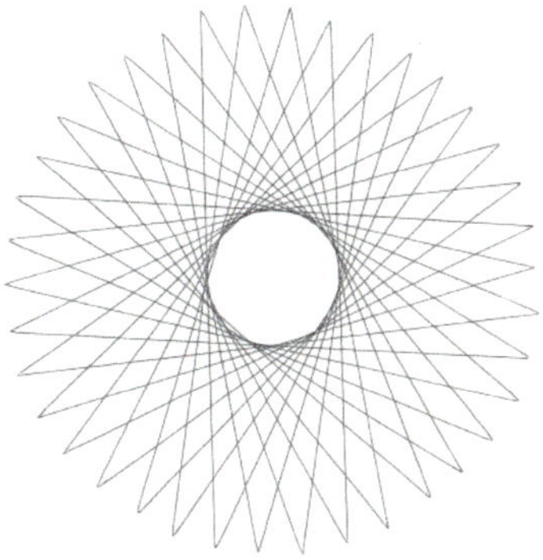

Sunburst

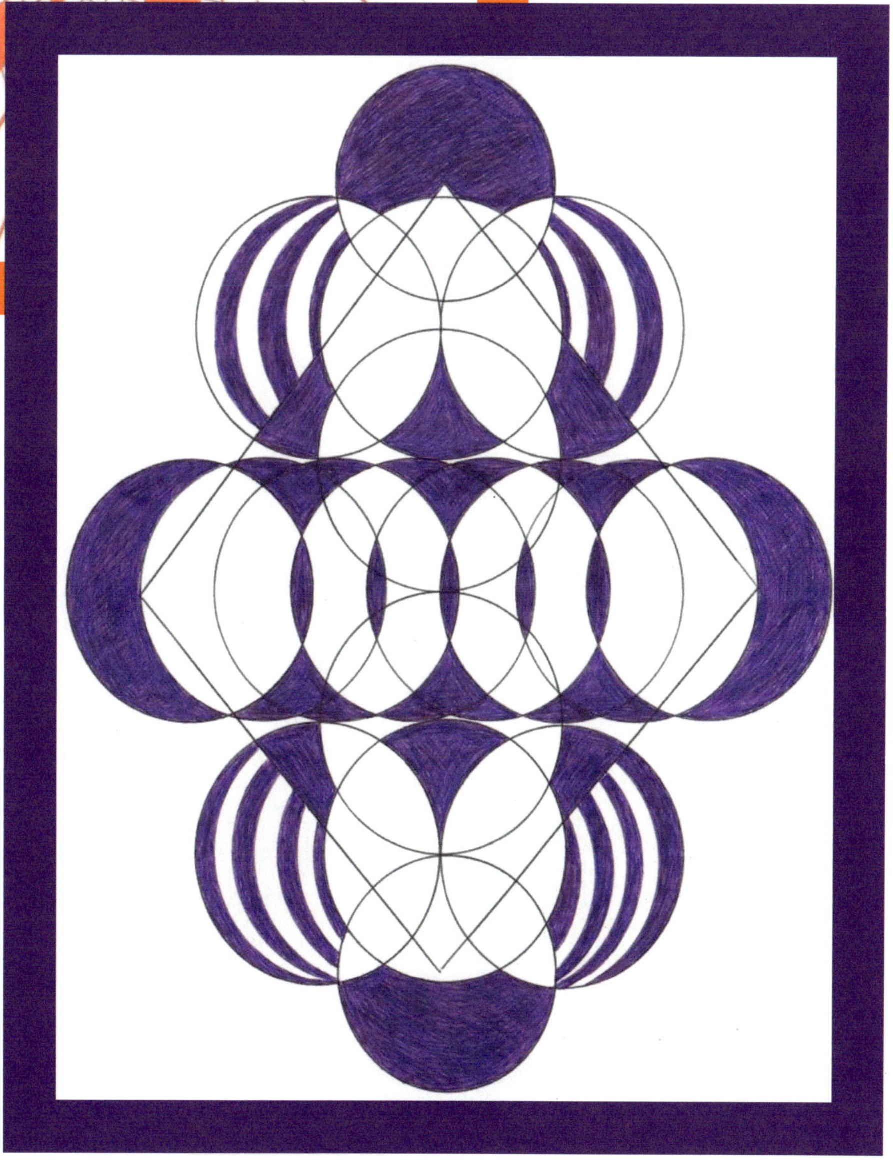

Circle Fantasy

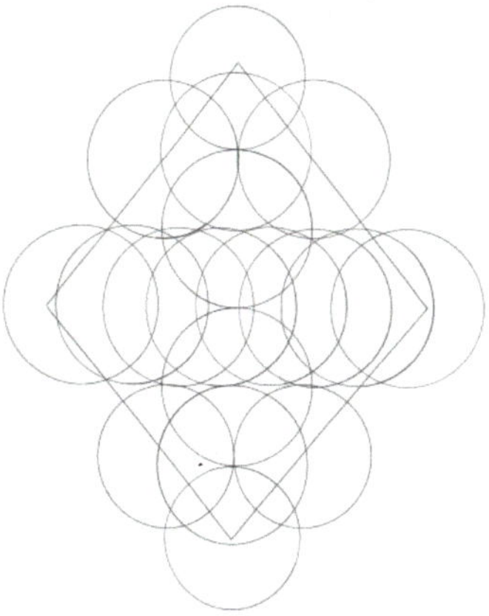

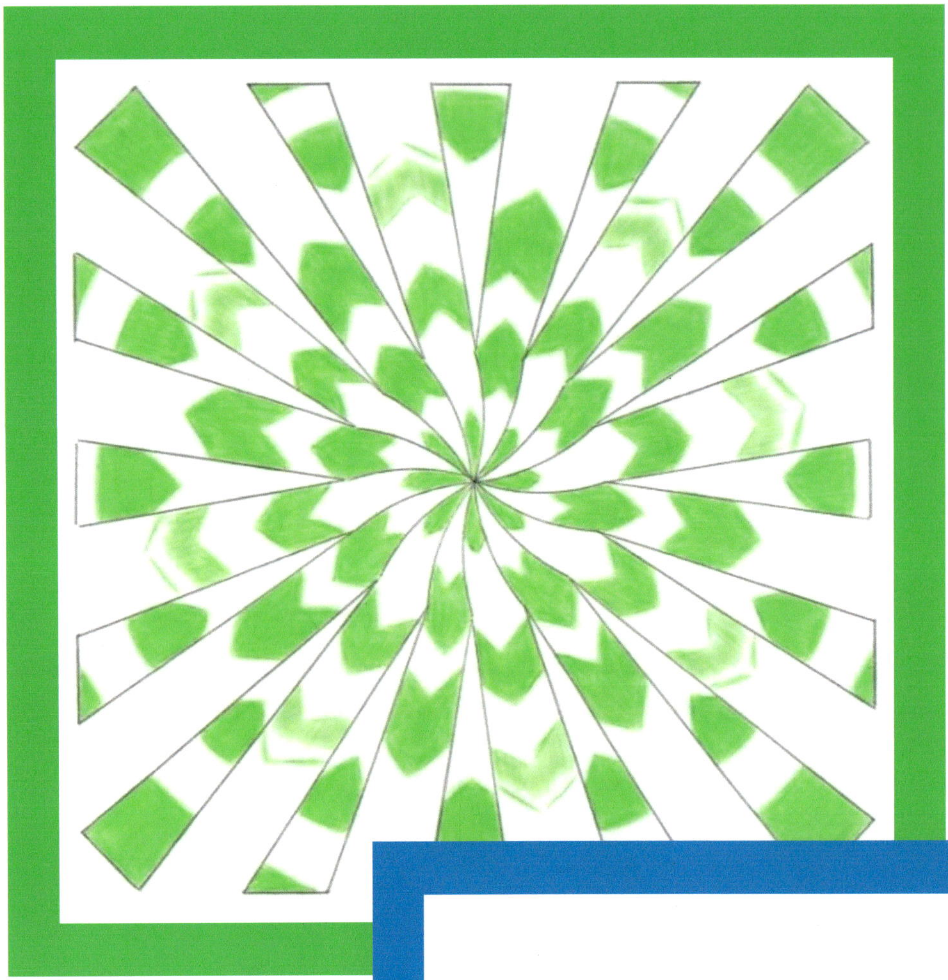

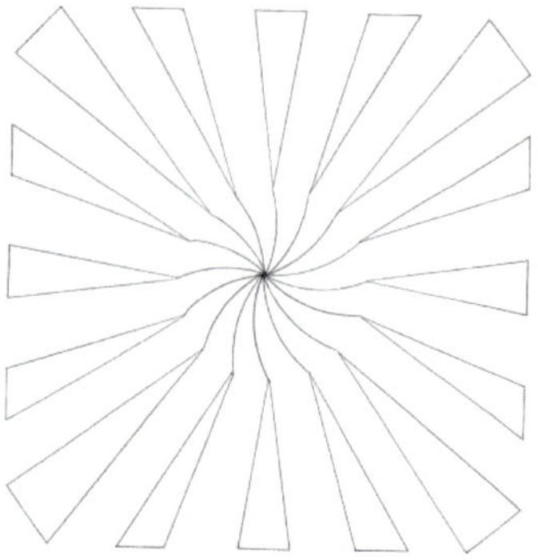

Green Sprouts

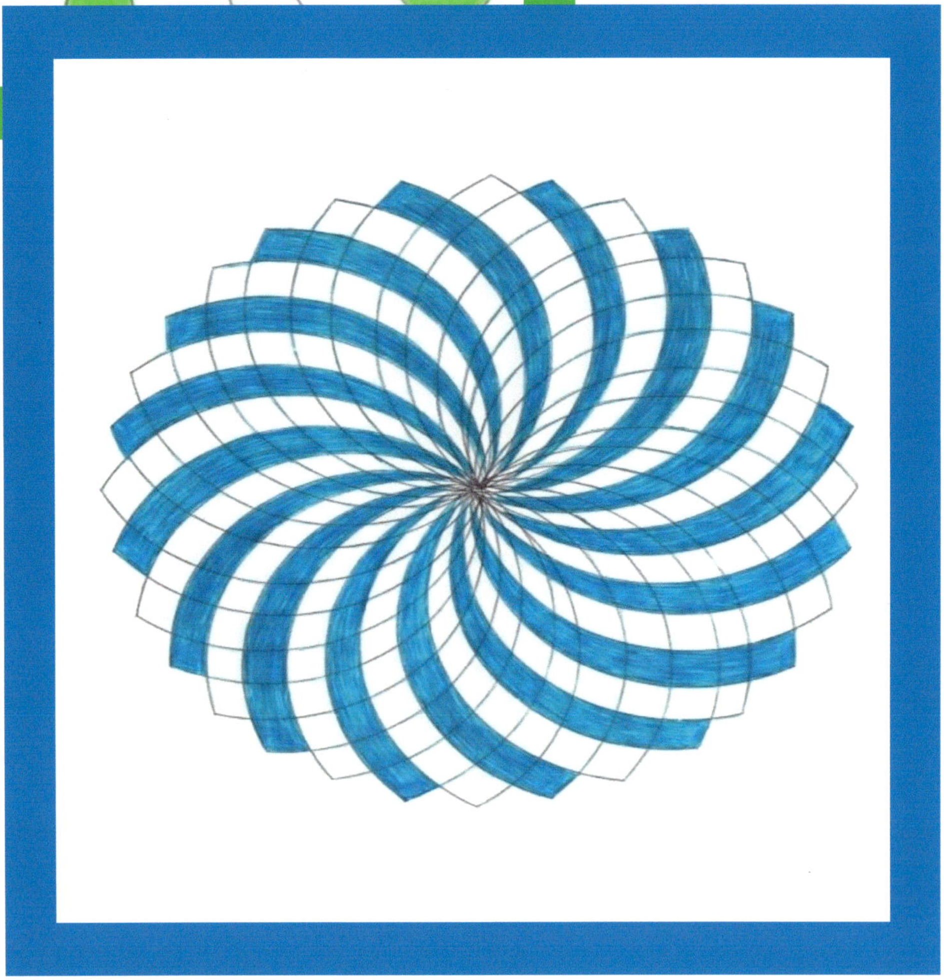

Splendid

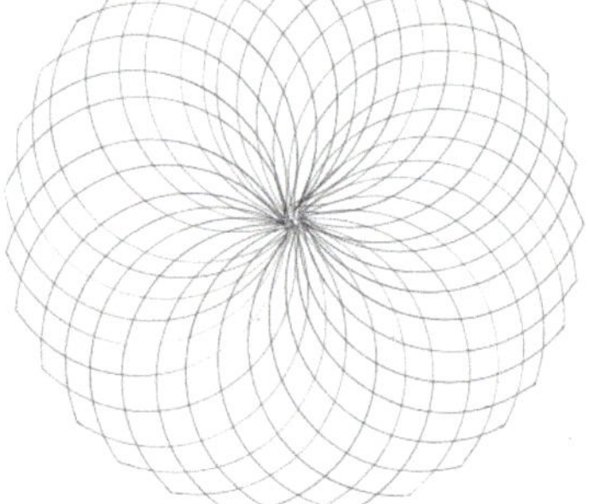

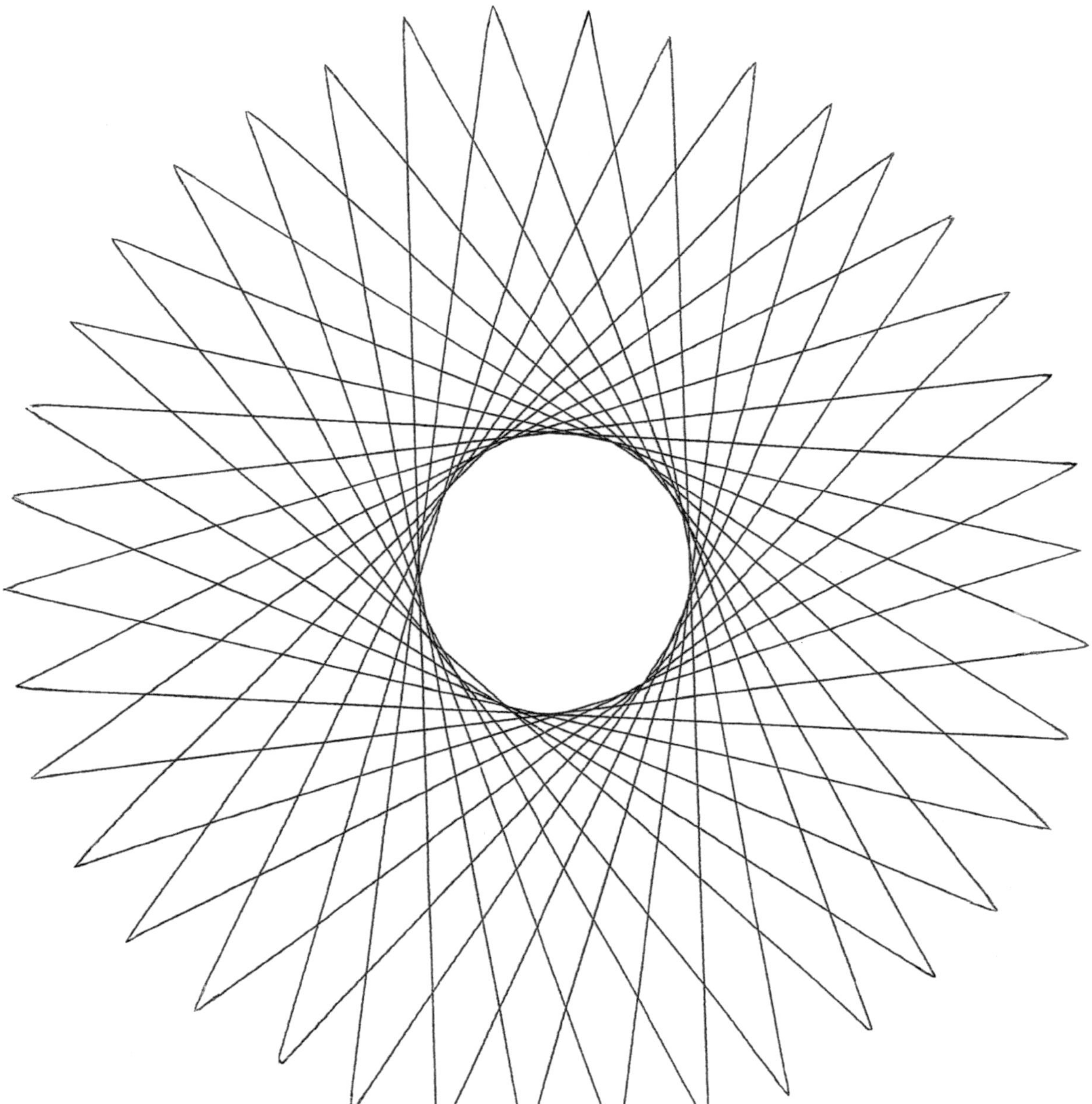

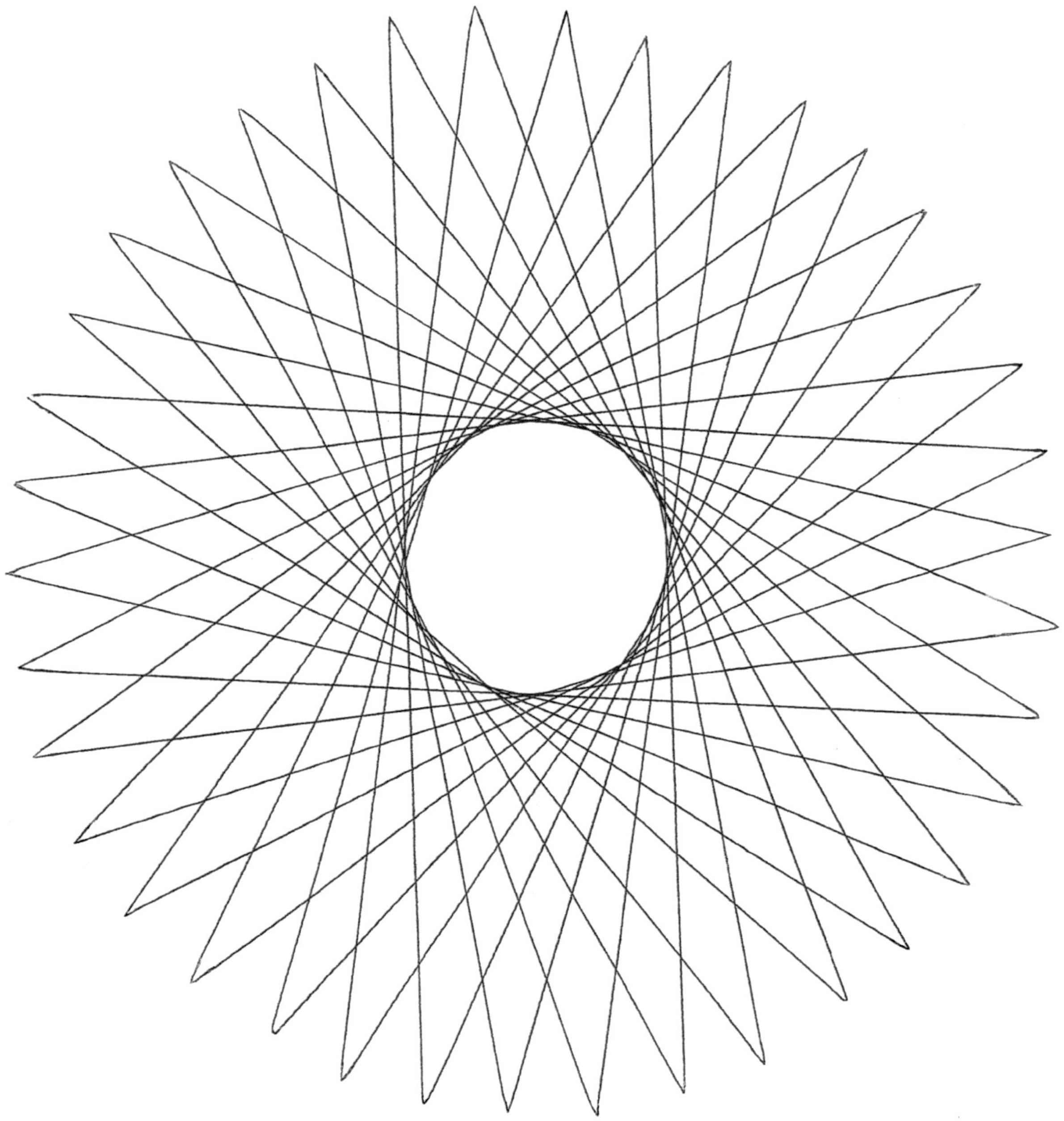

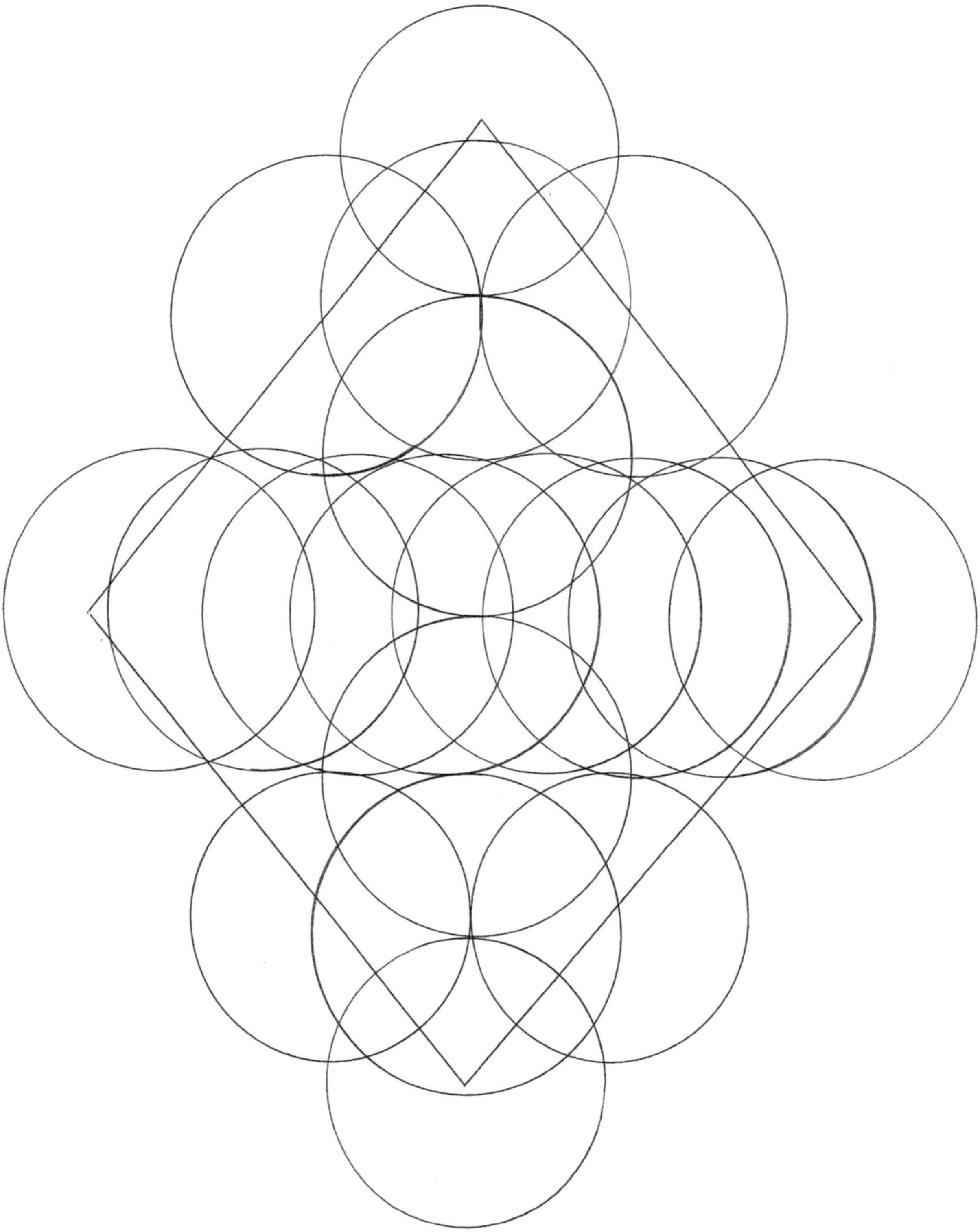

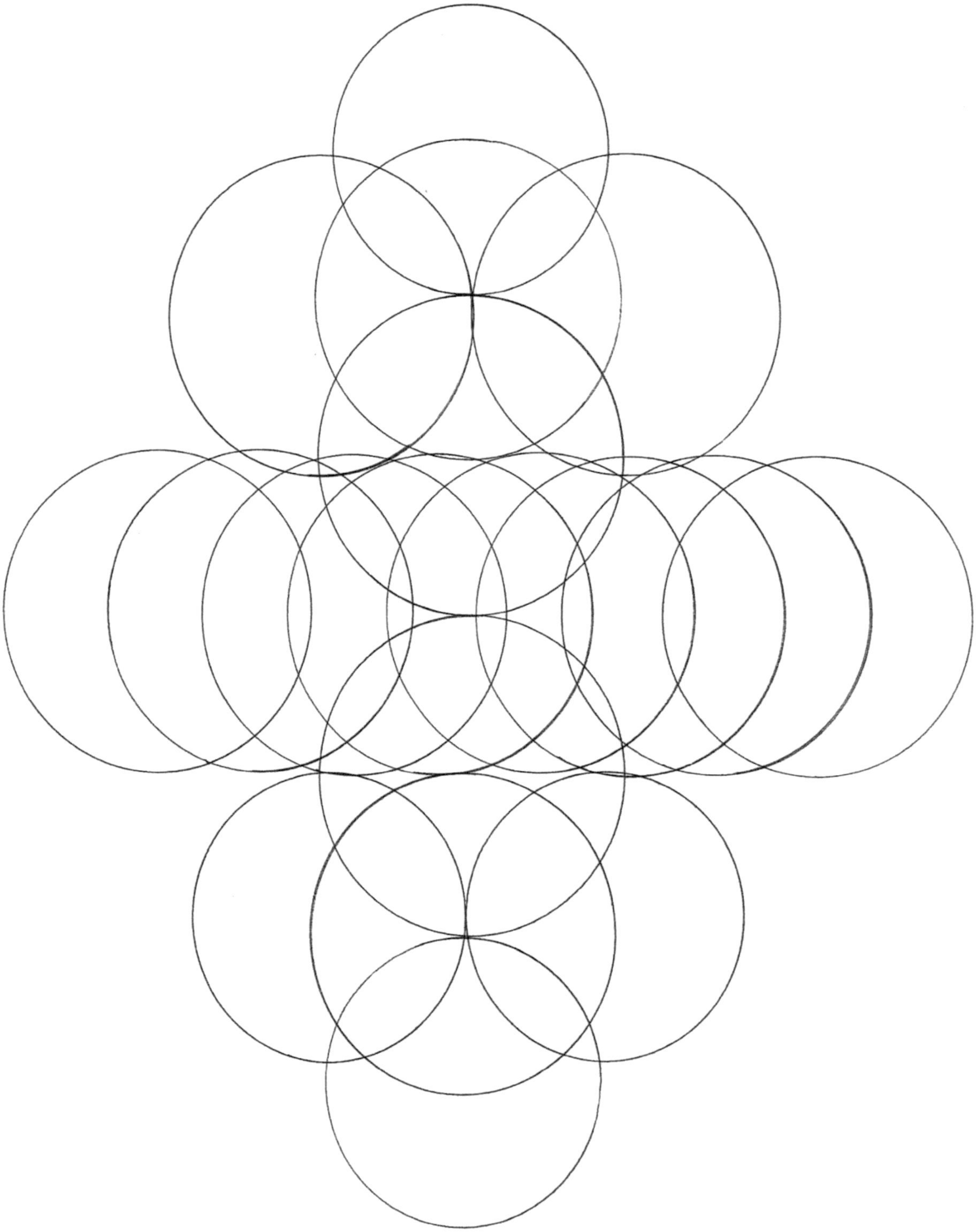

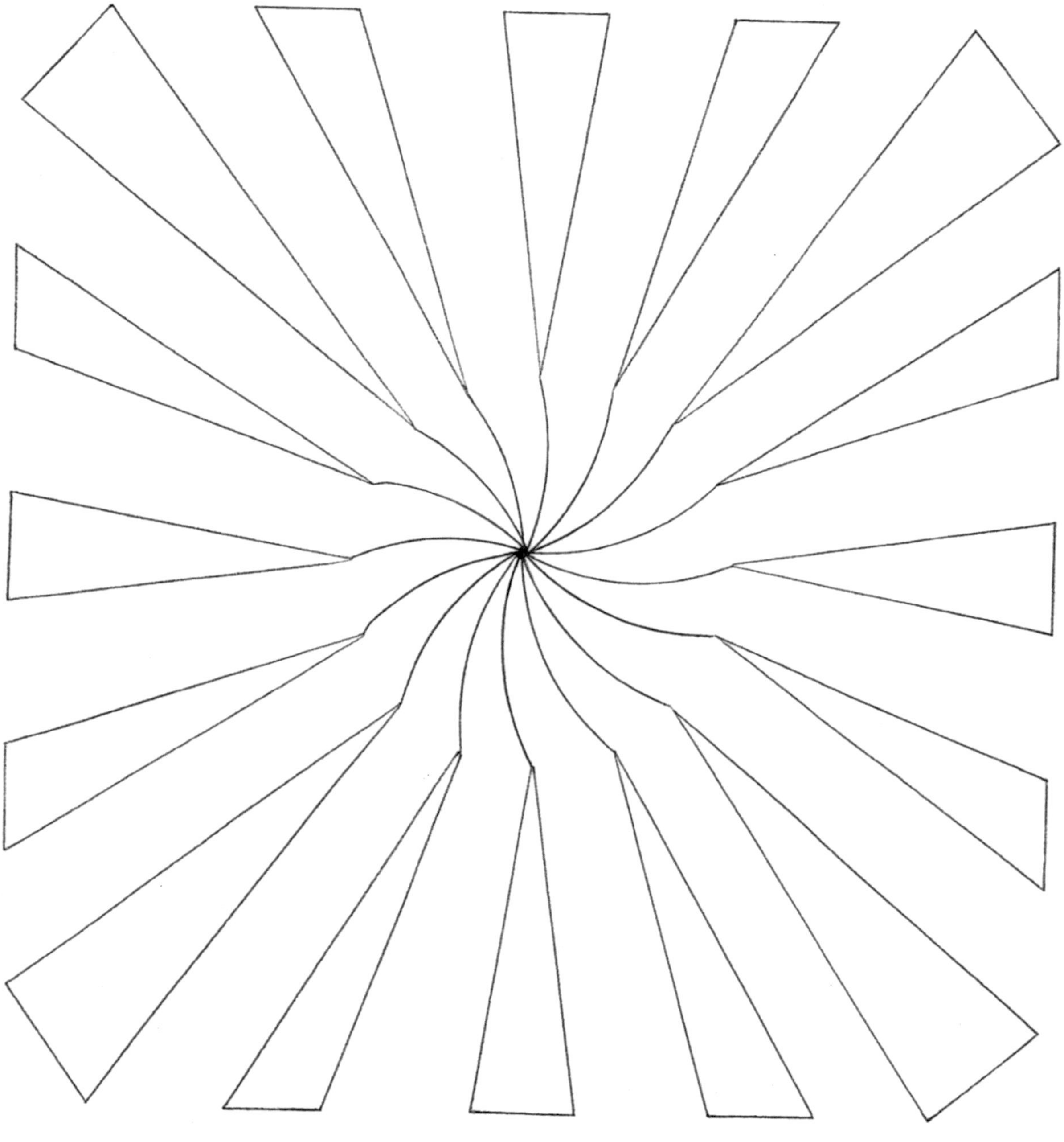

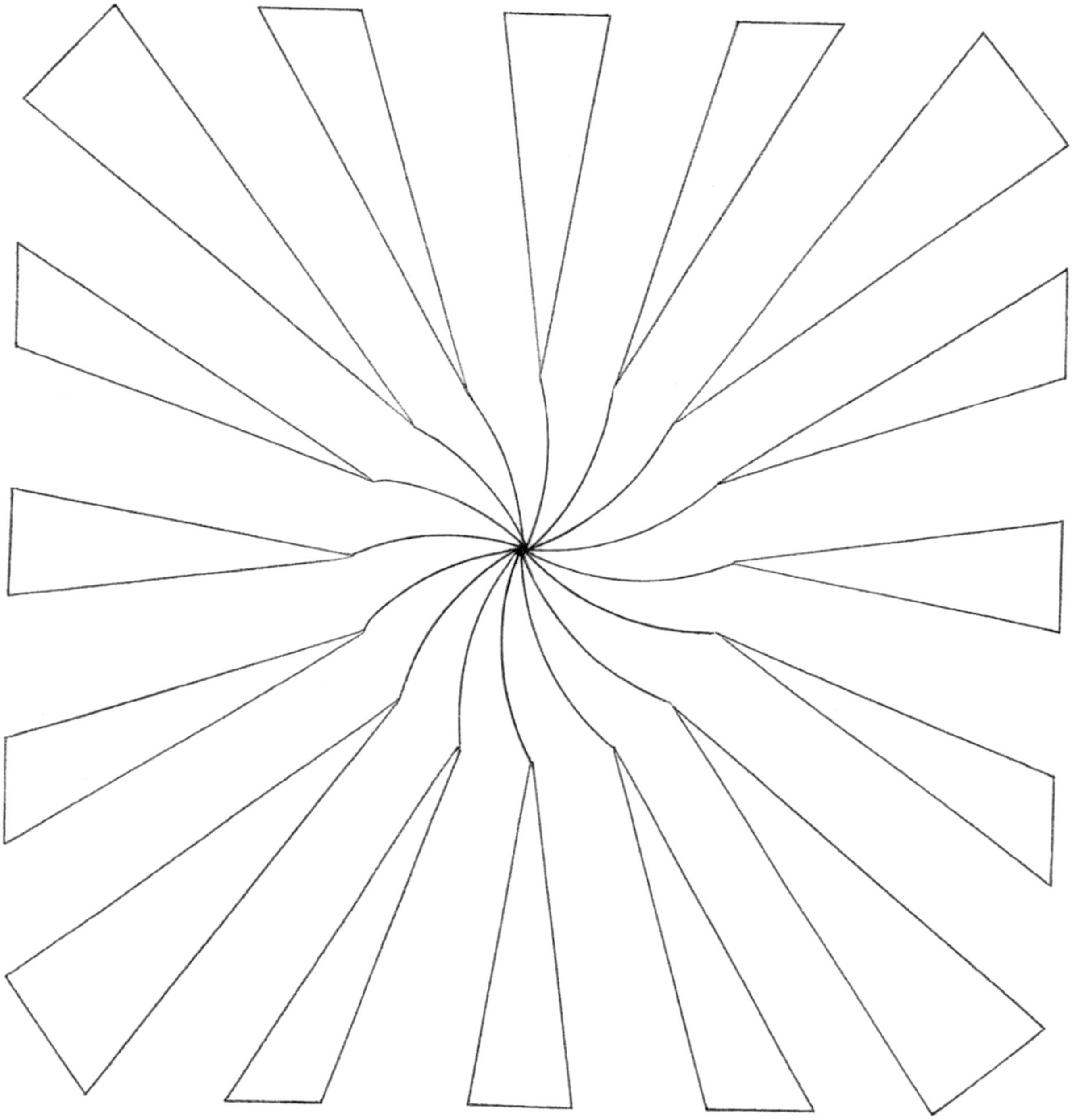

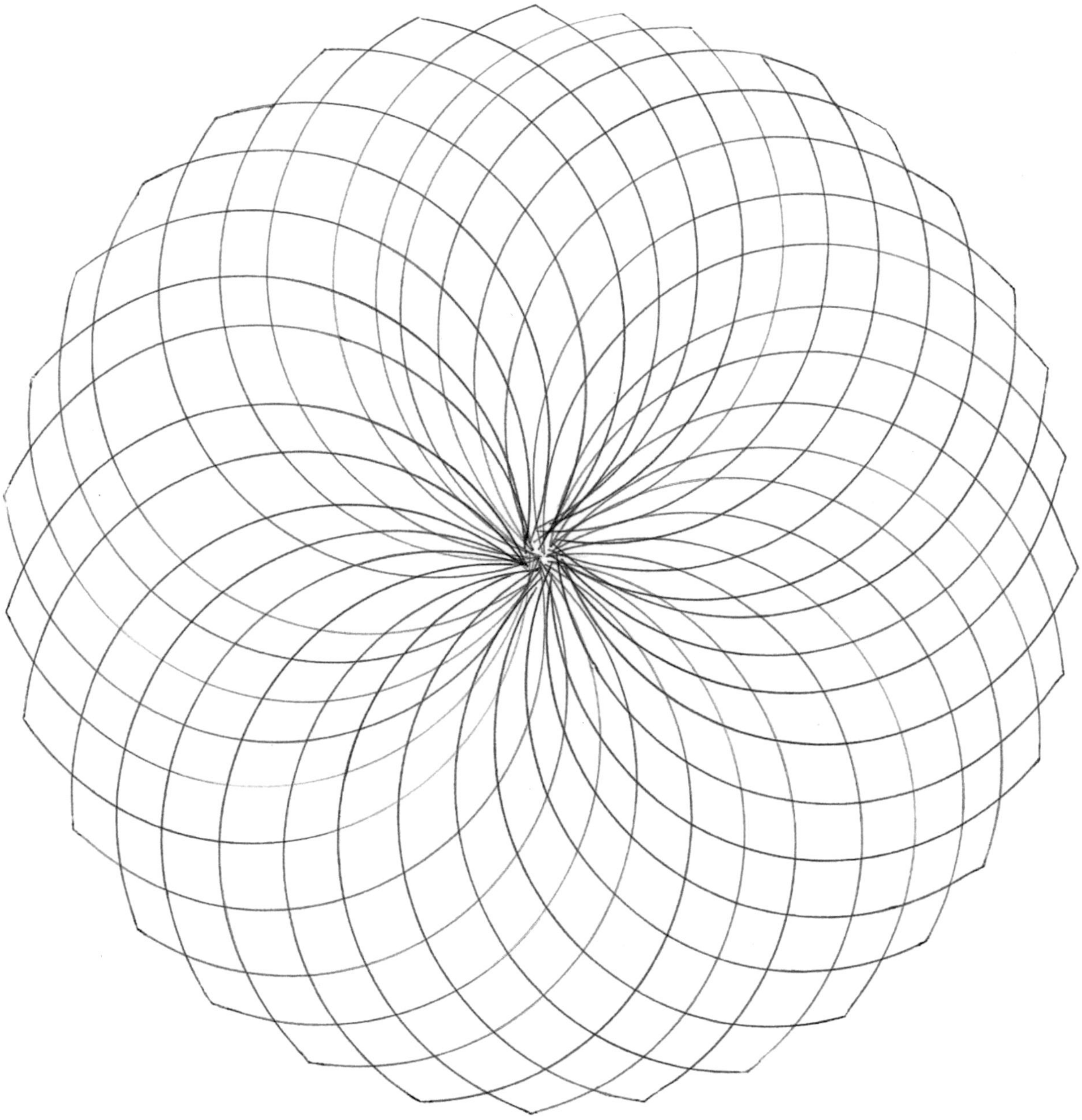

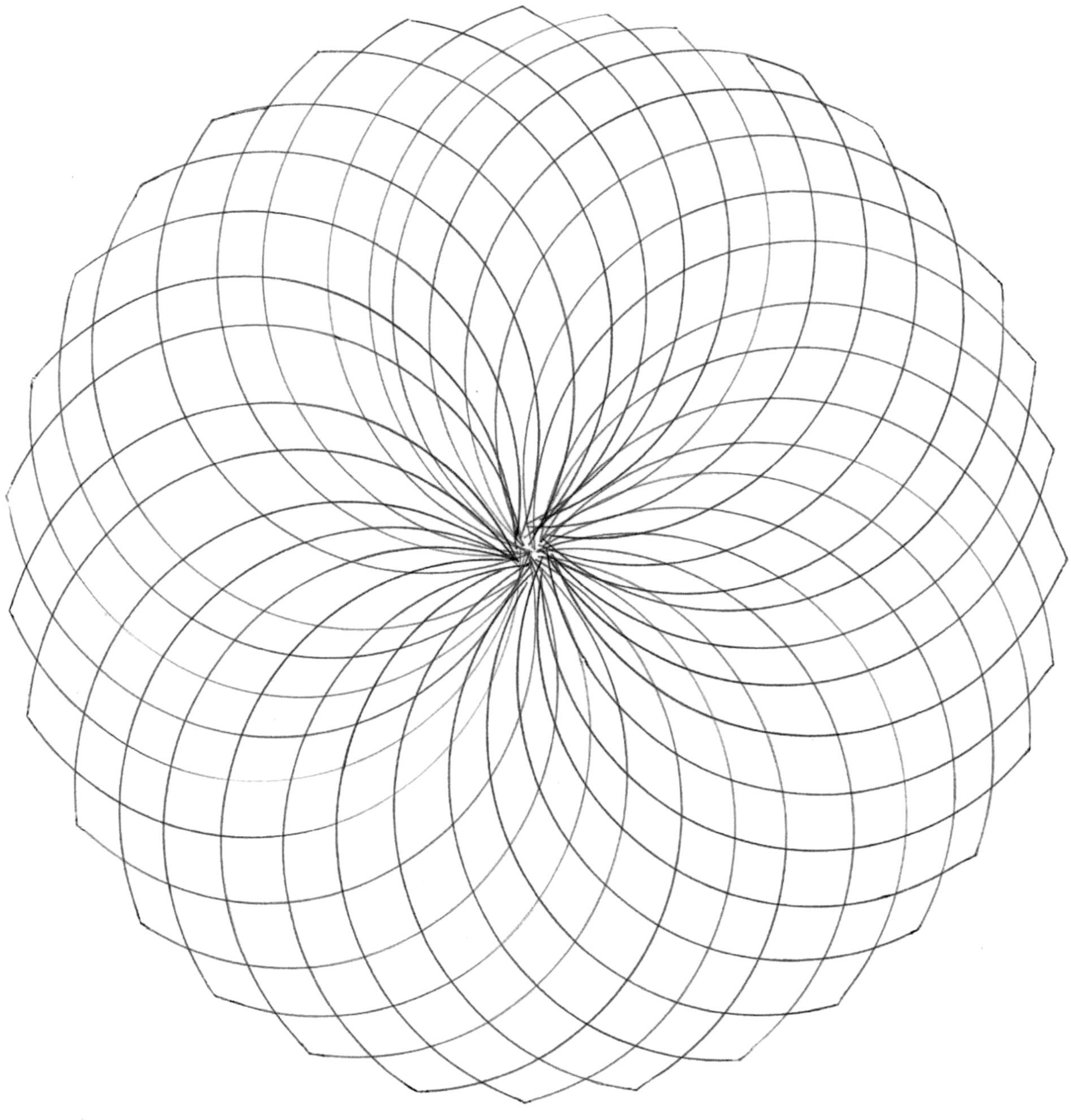

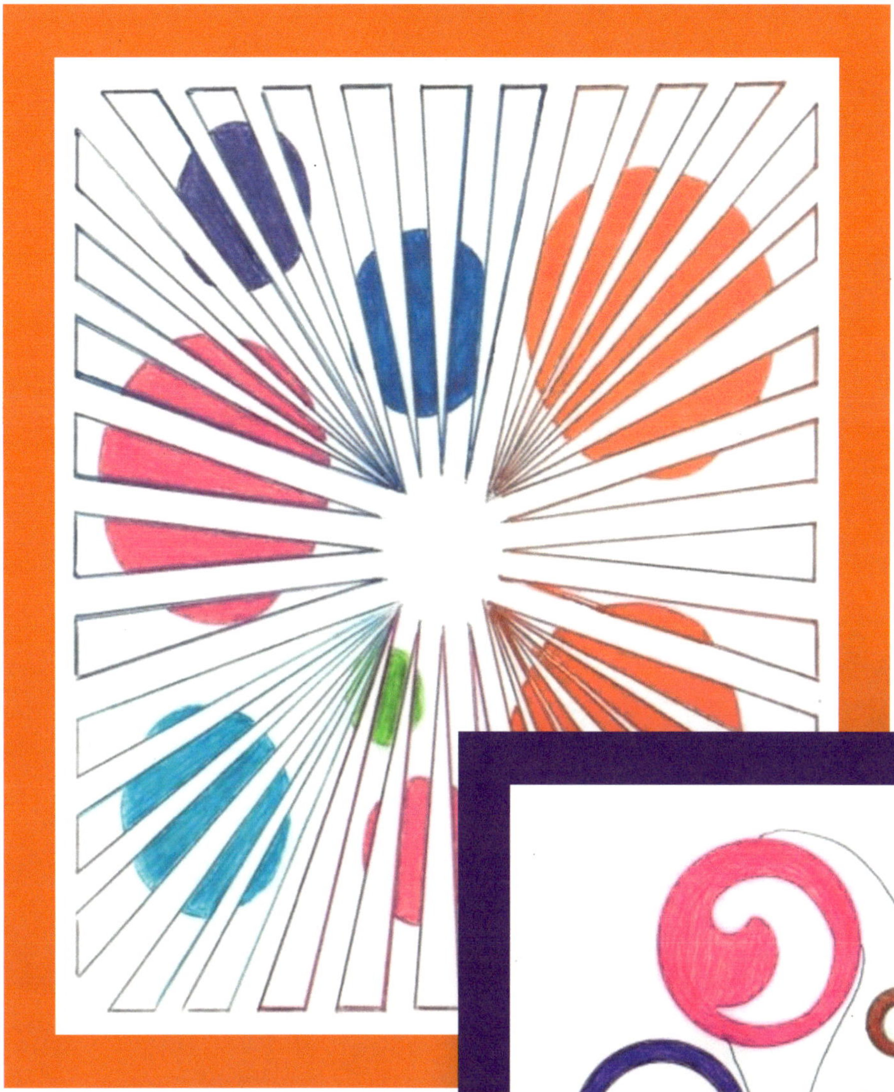

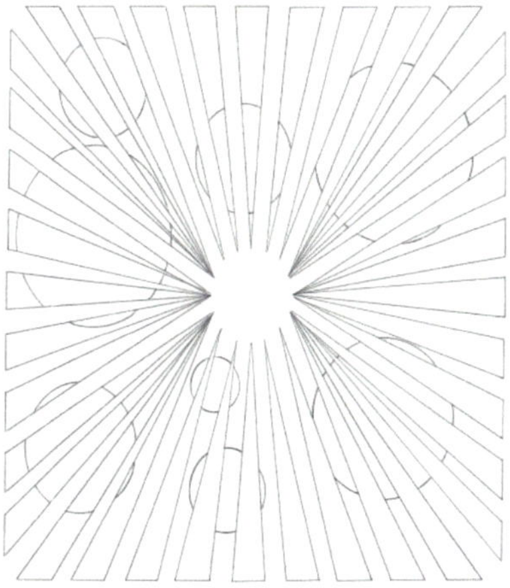

Hidden Shapes

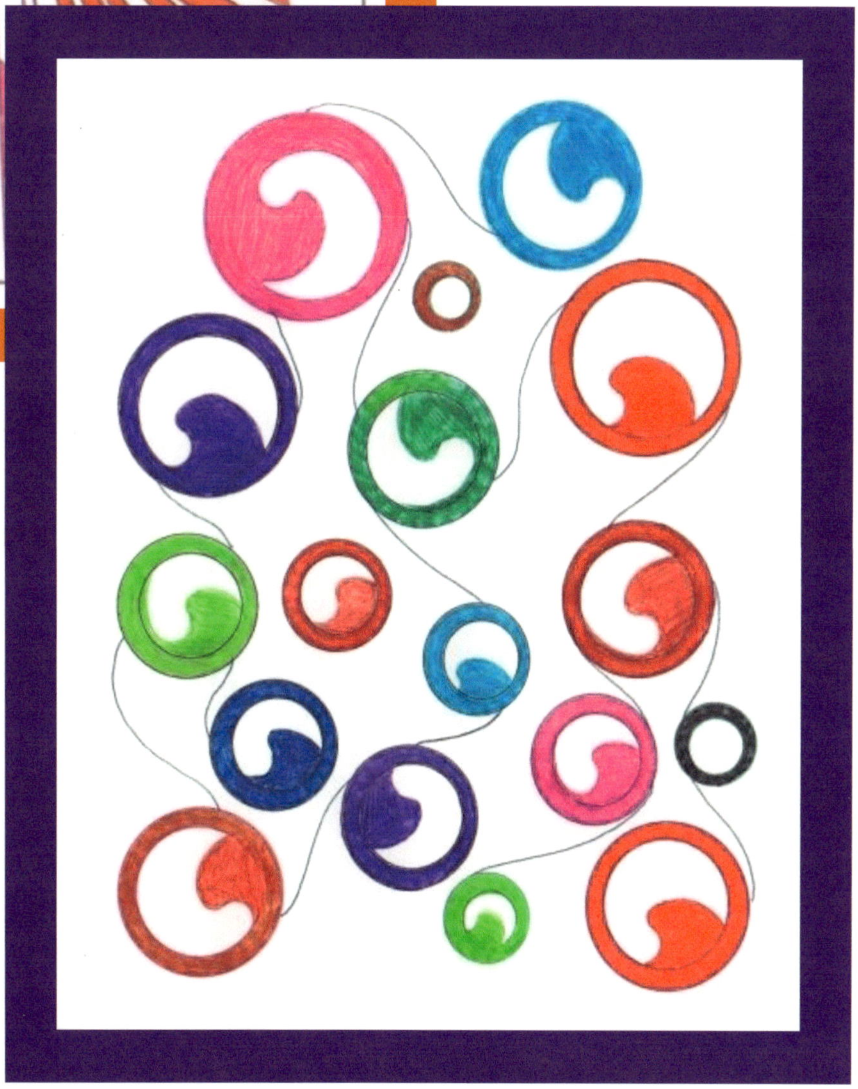

Circles Galore

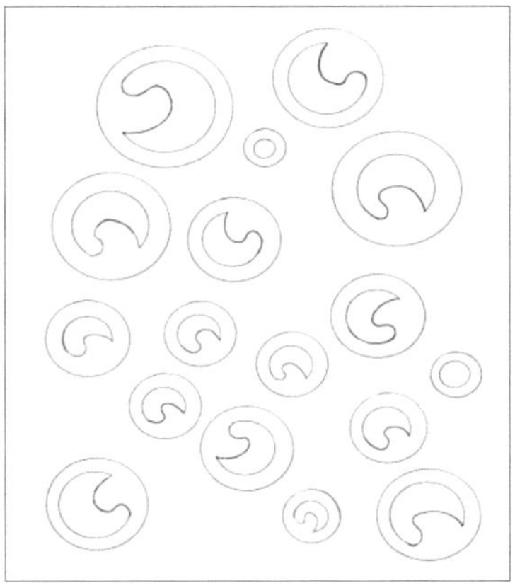

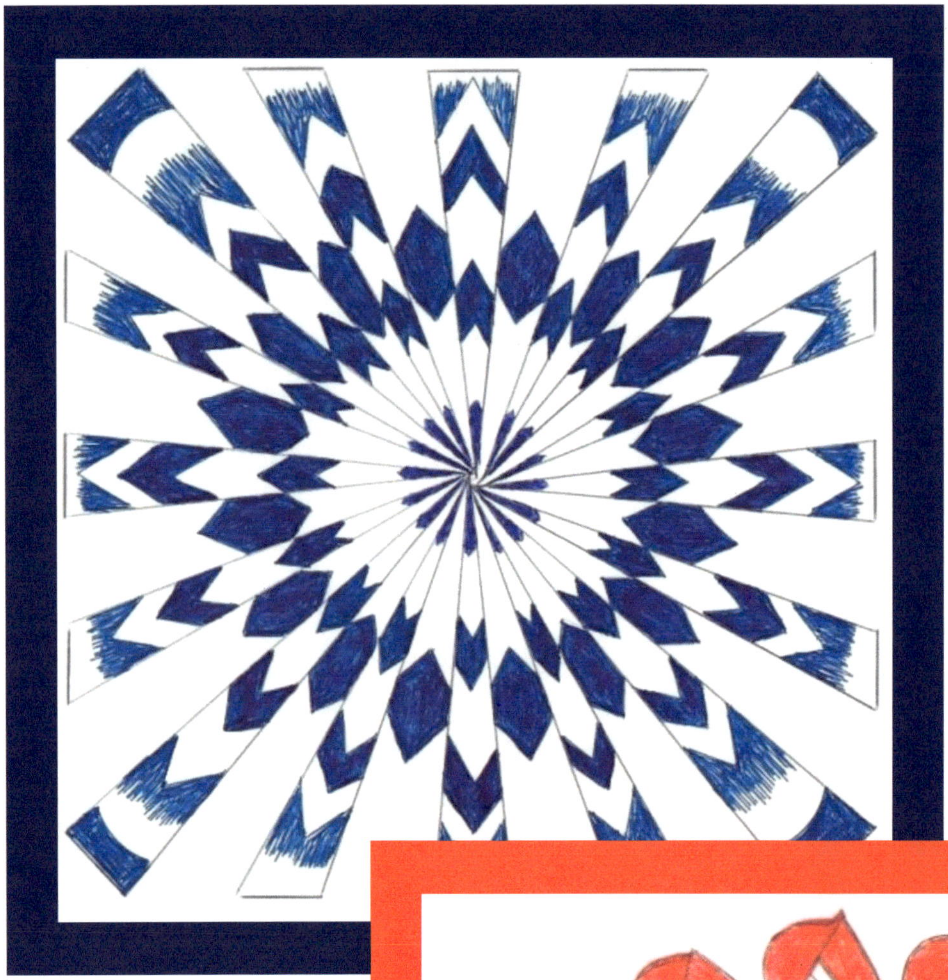

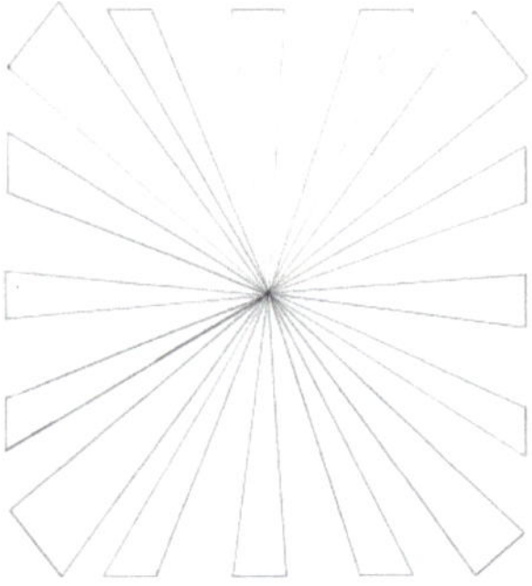

Rhapsody in Blue

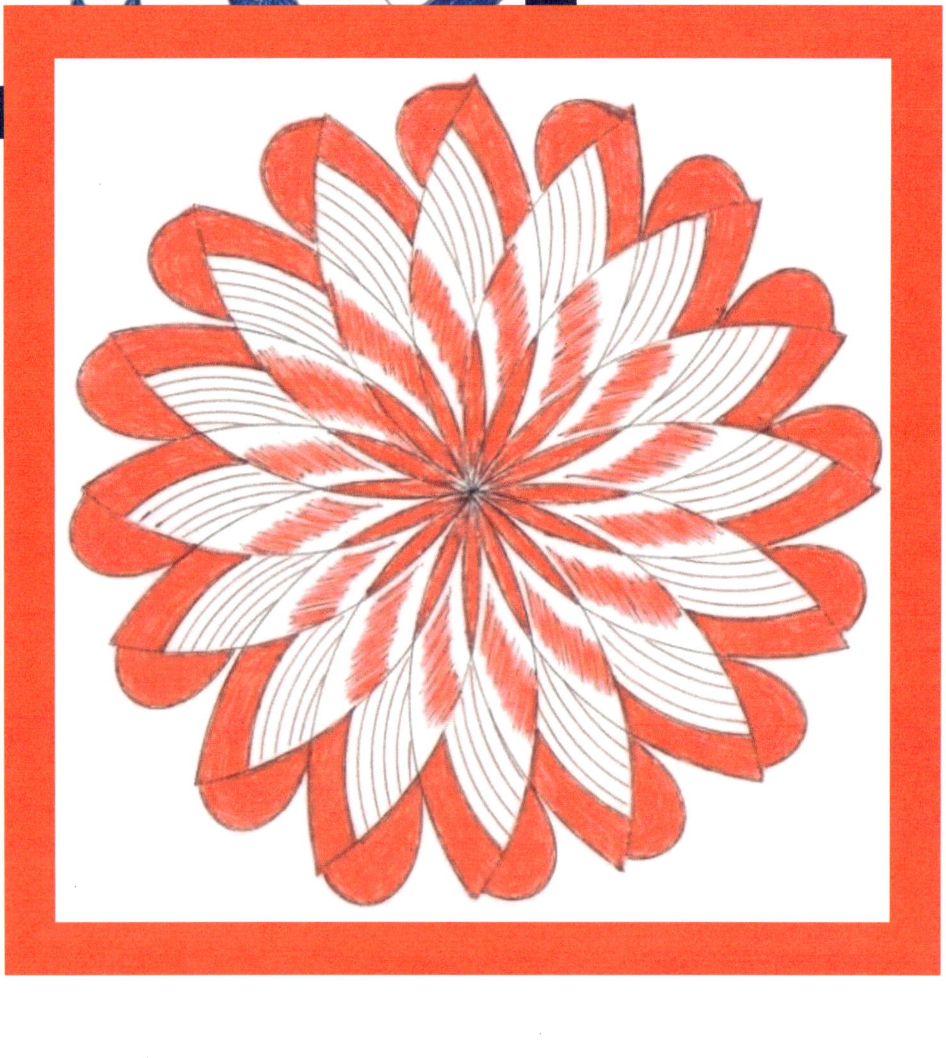

Incredible

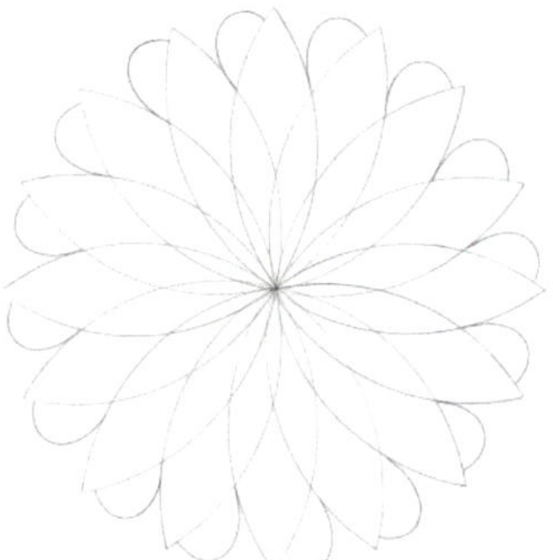

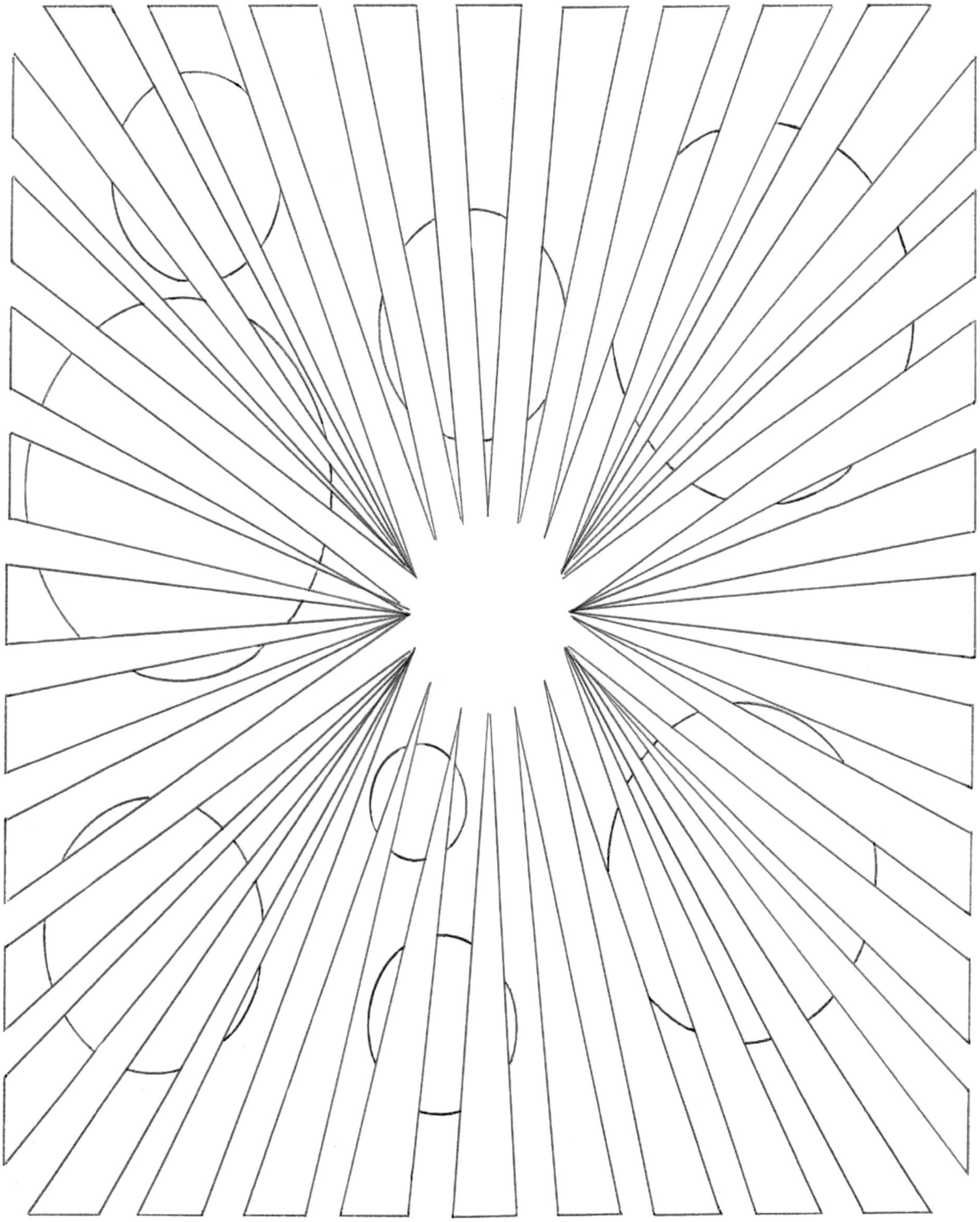

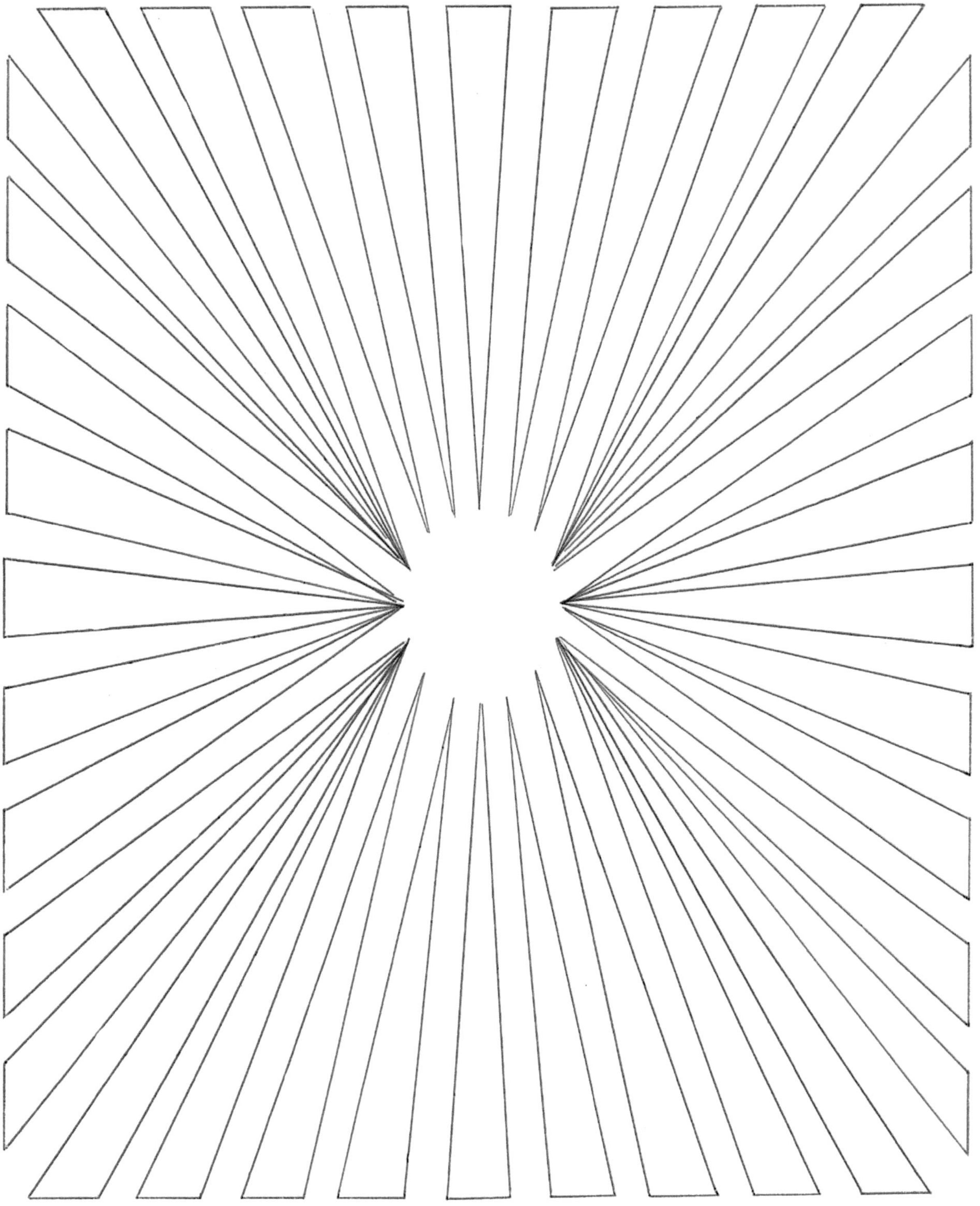

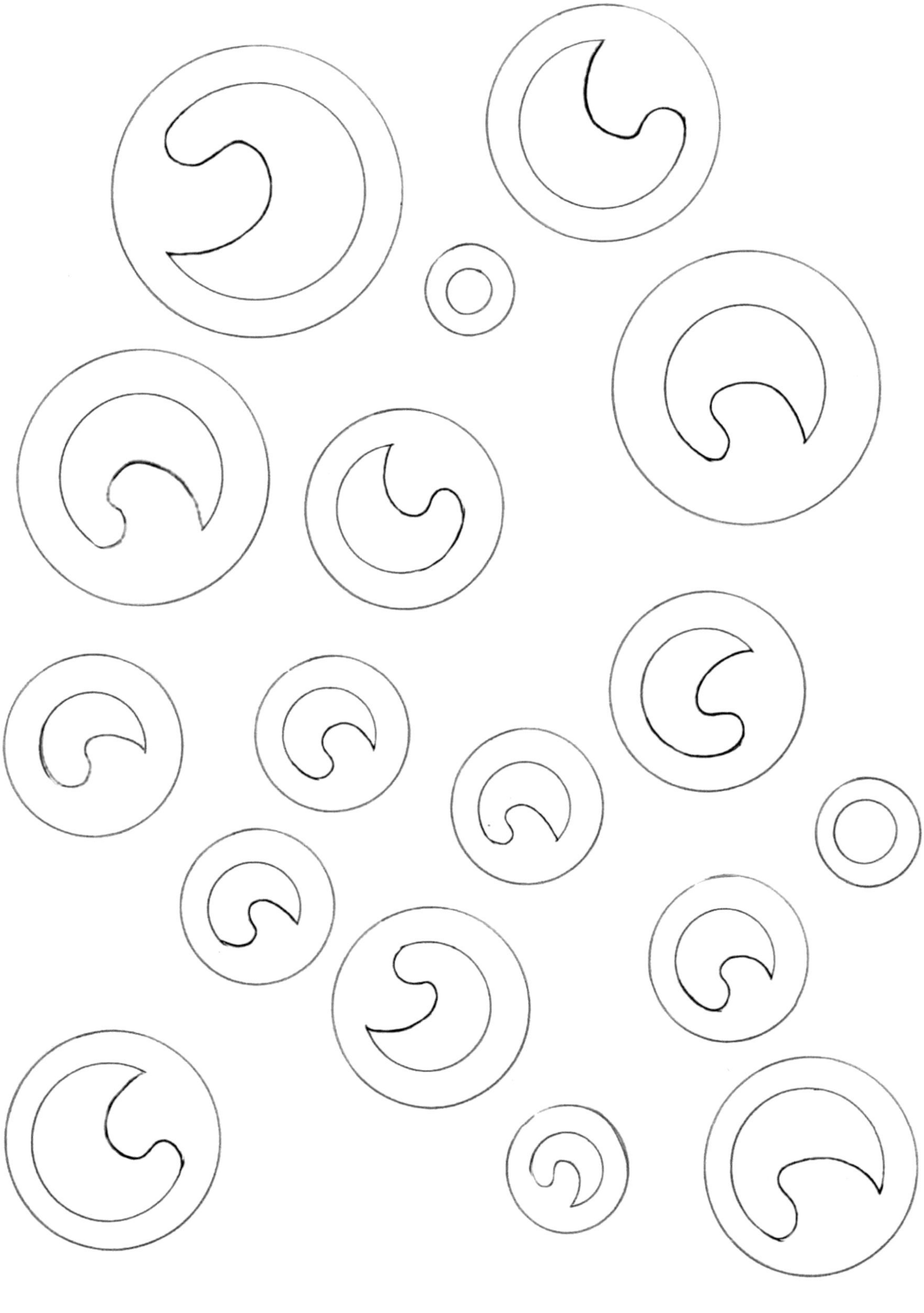

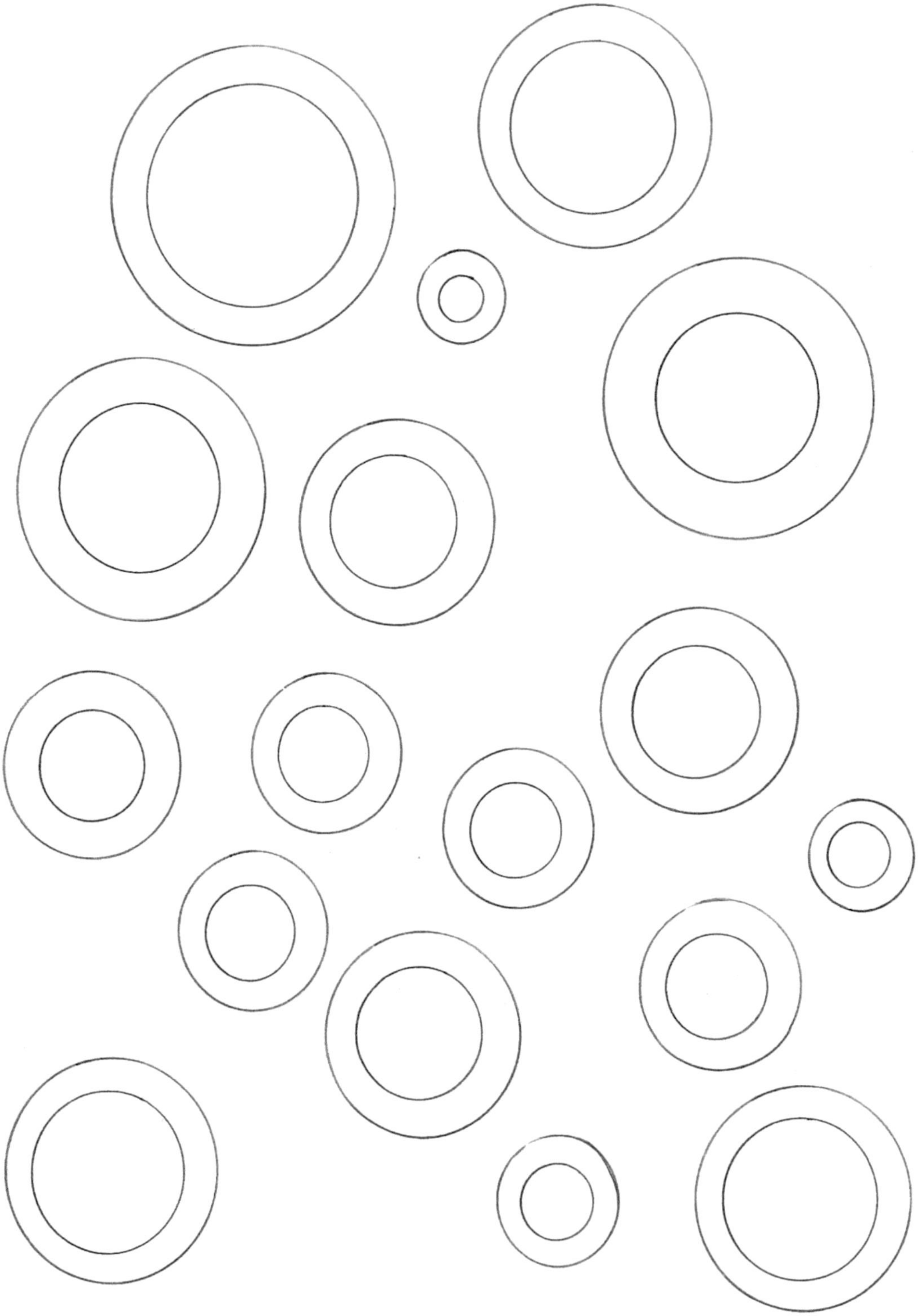

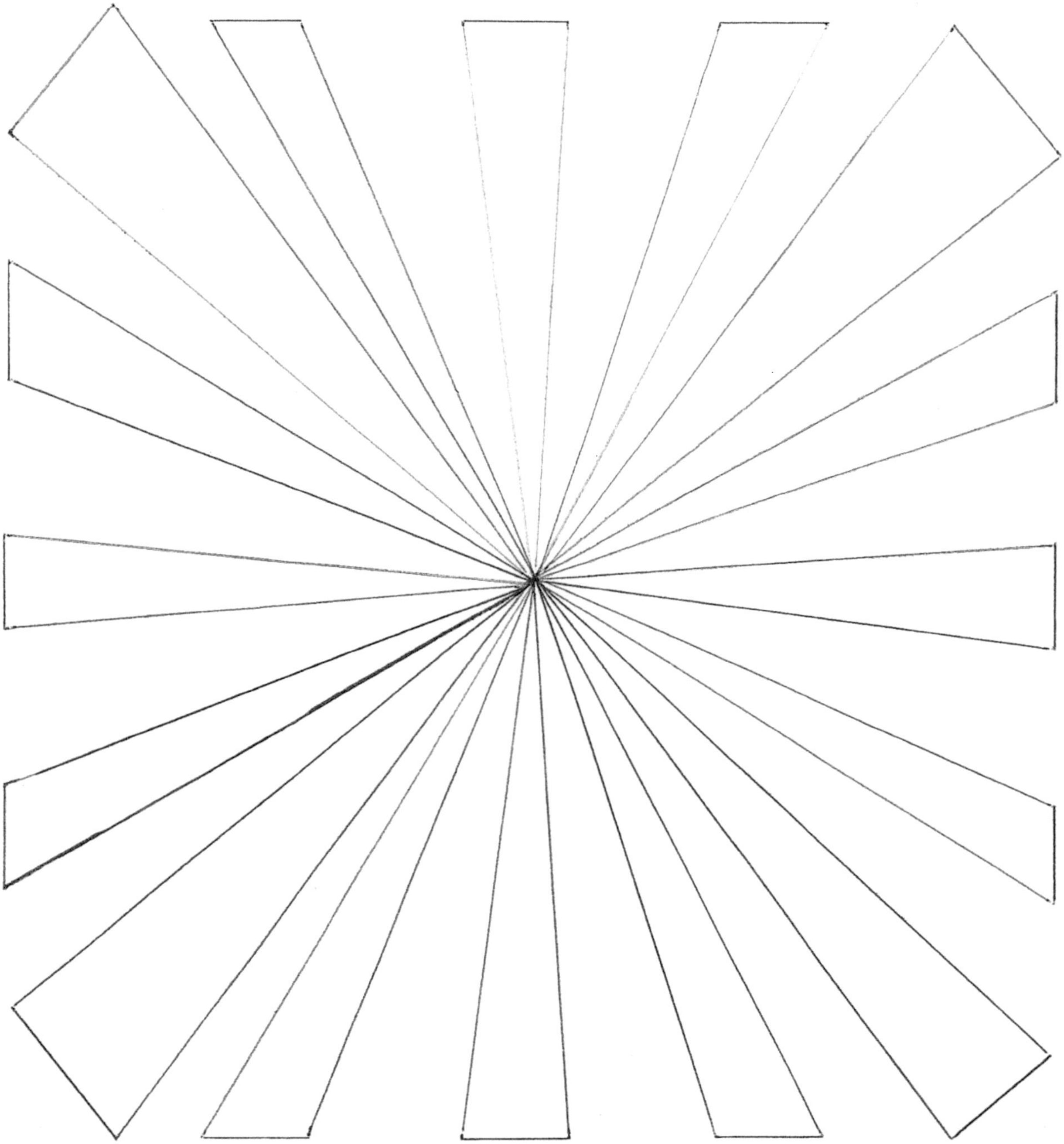

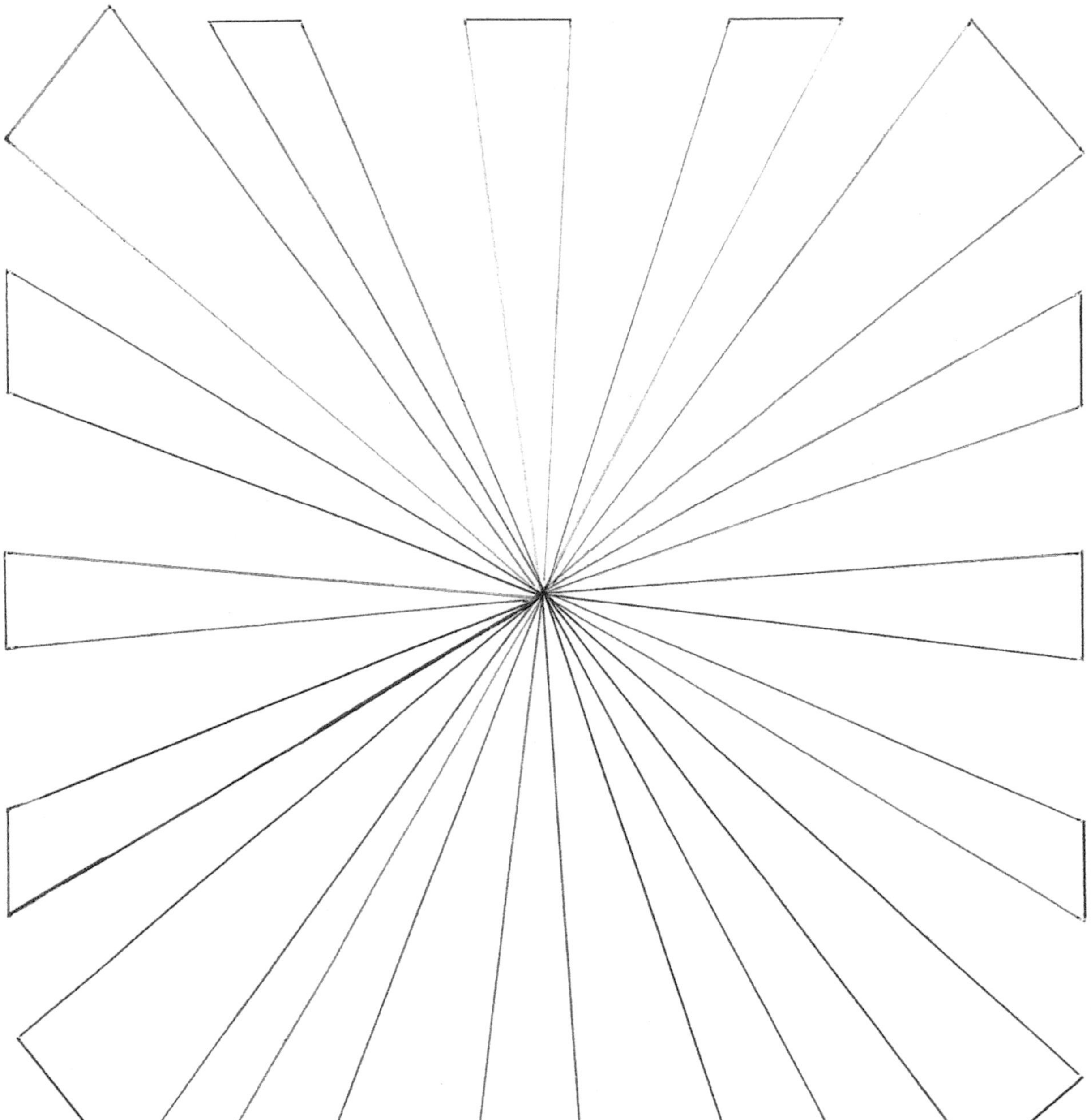

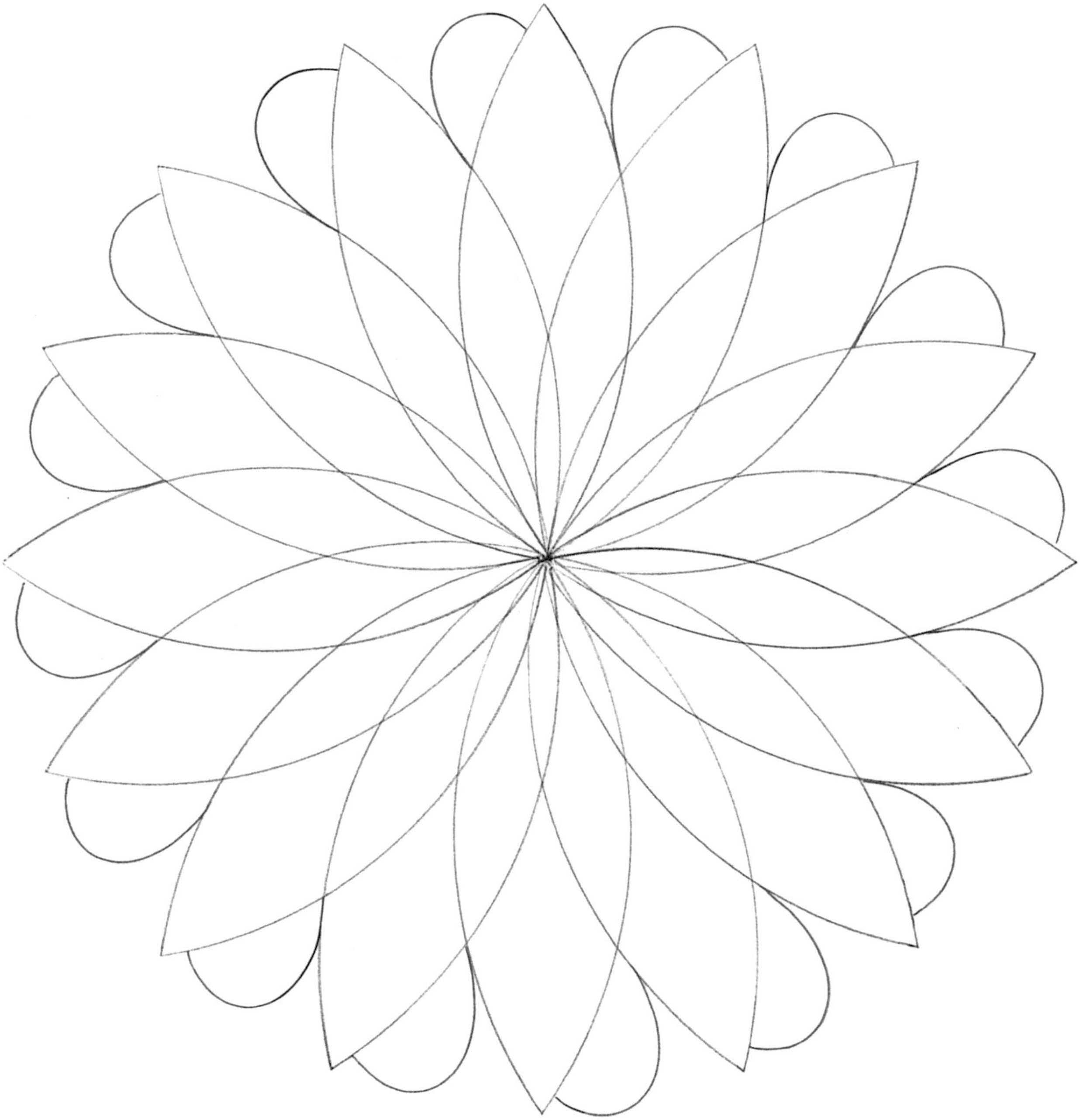

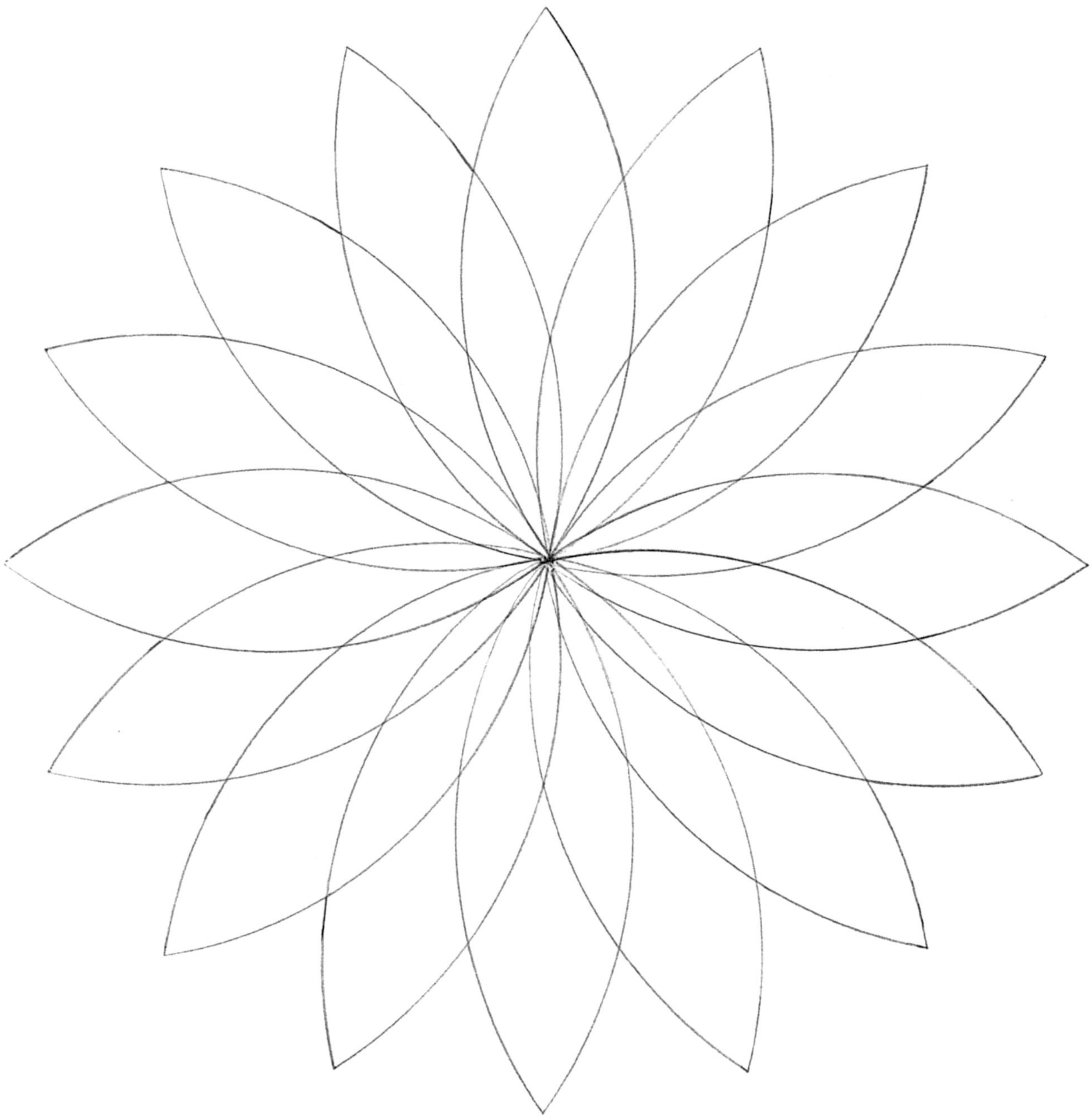

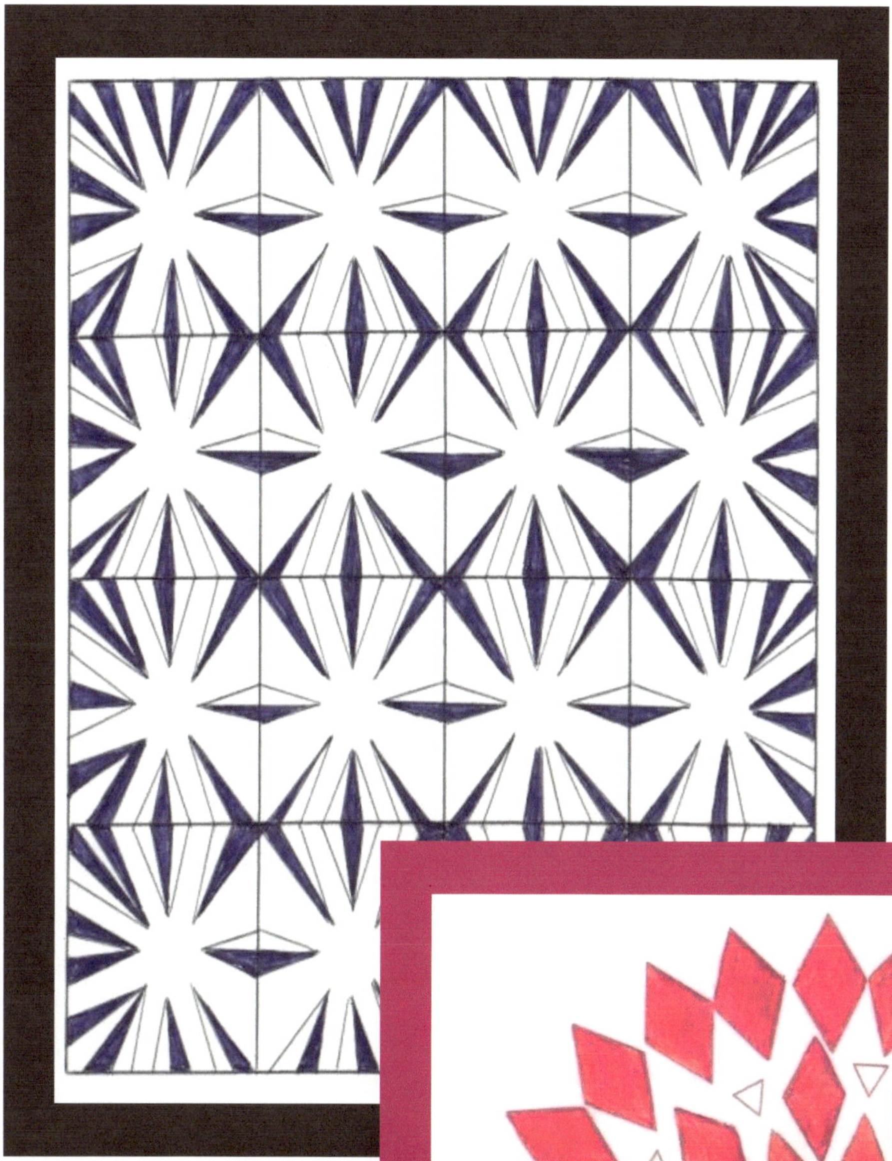

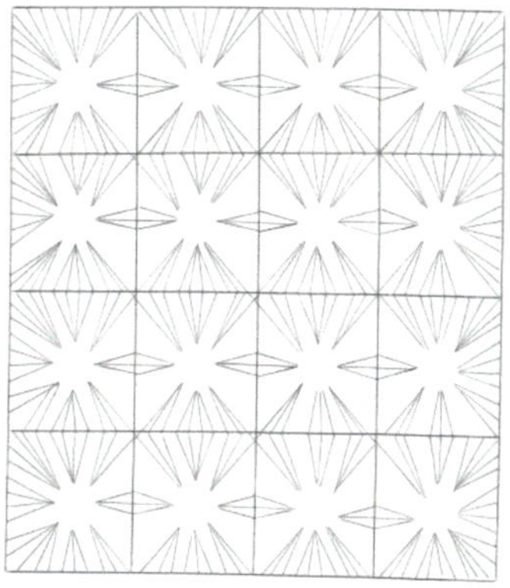

Wallpaper

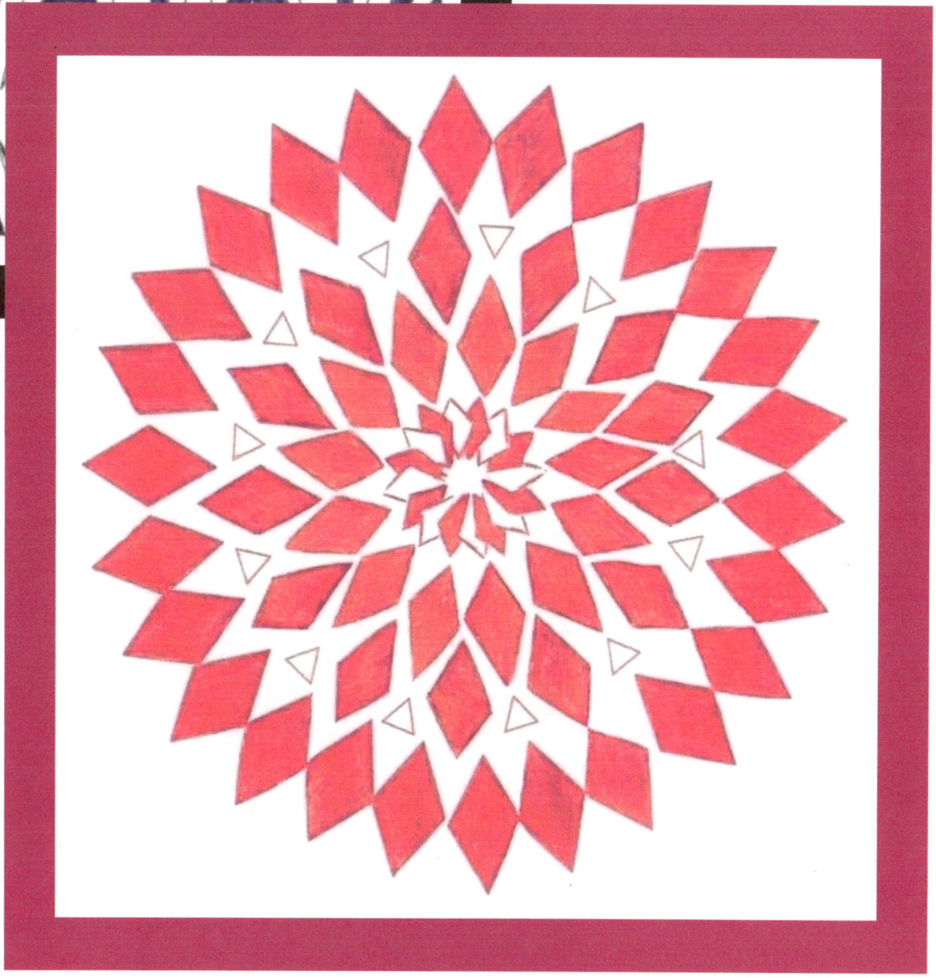

Diamonds are Forever

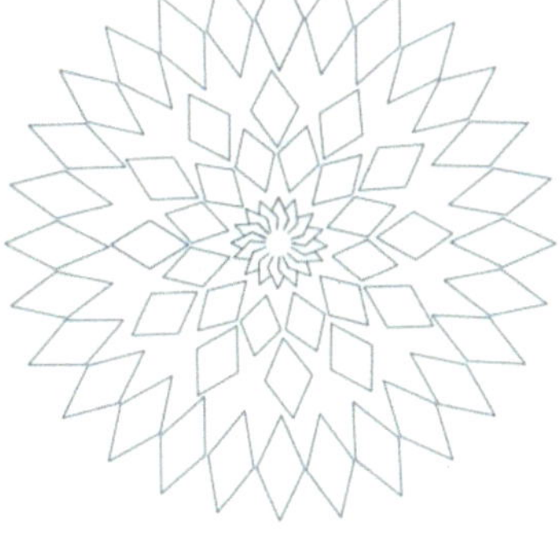

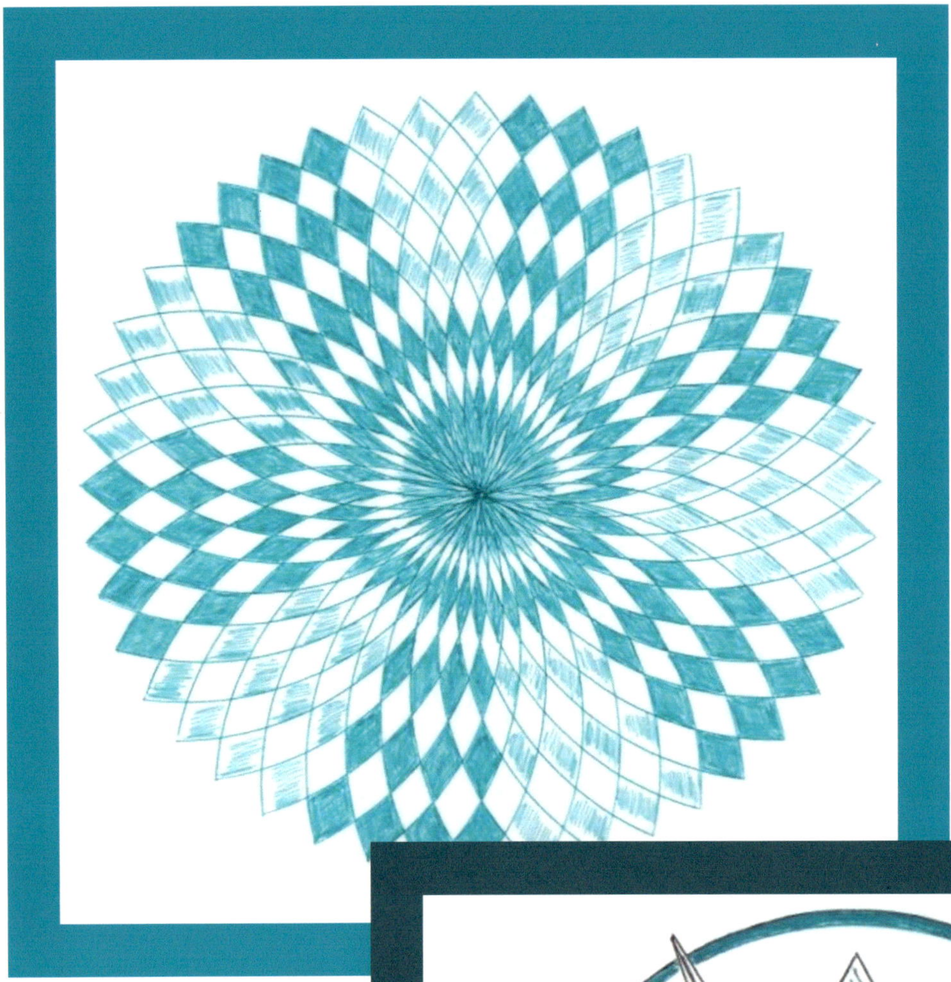

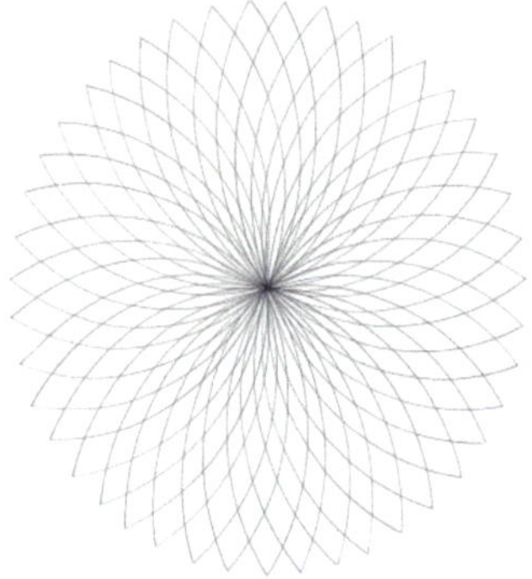

Capture the Beauty

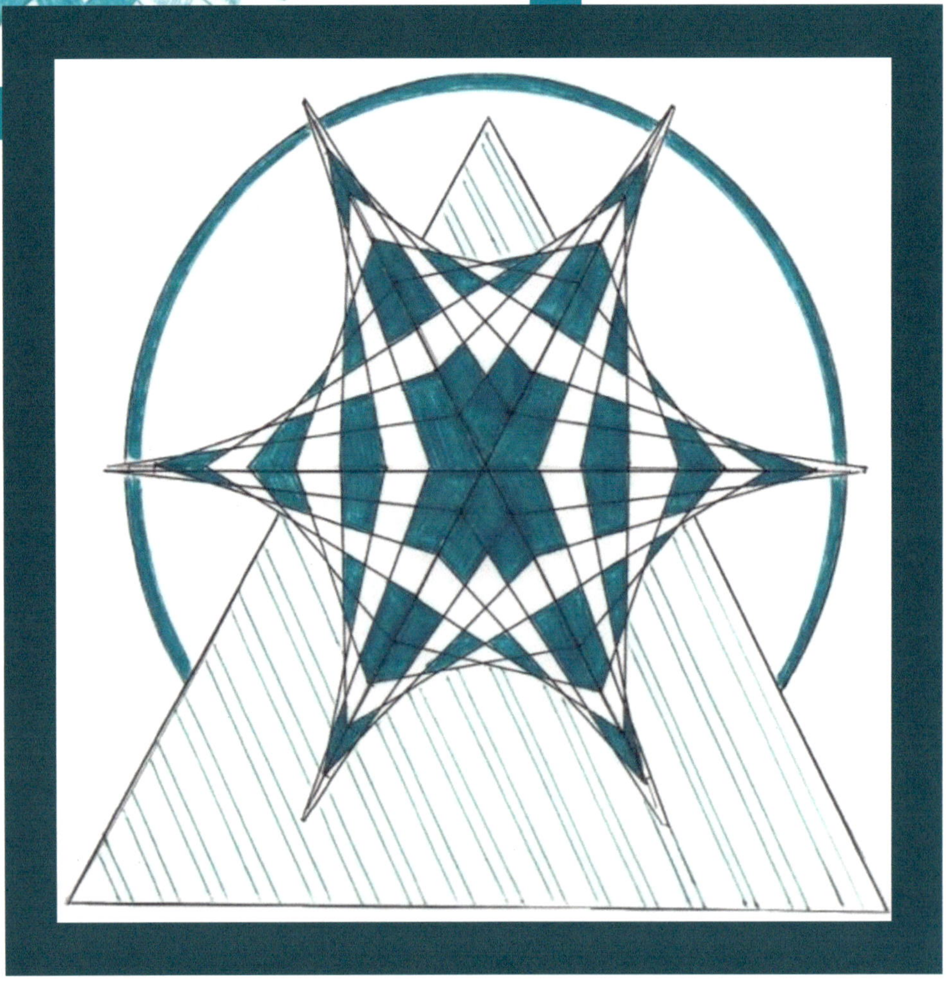

Beauty of Geometry

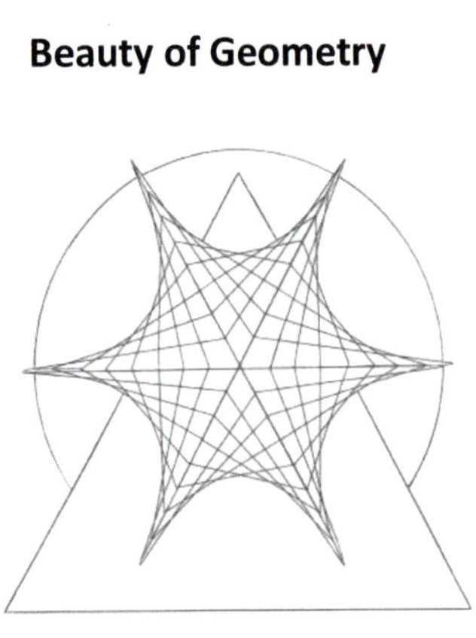

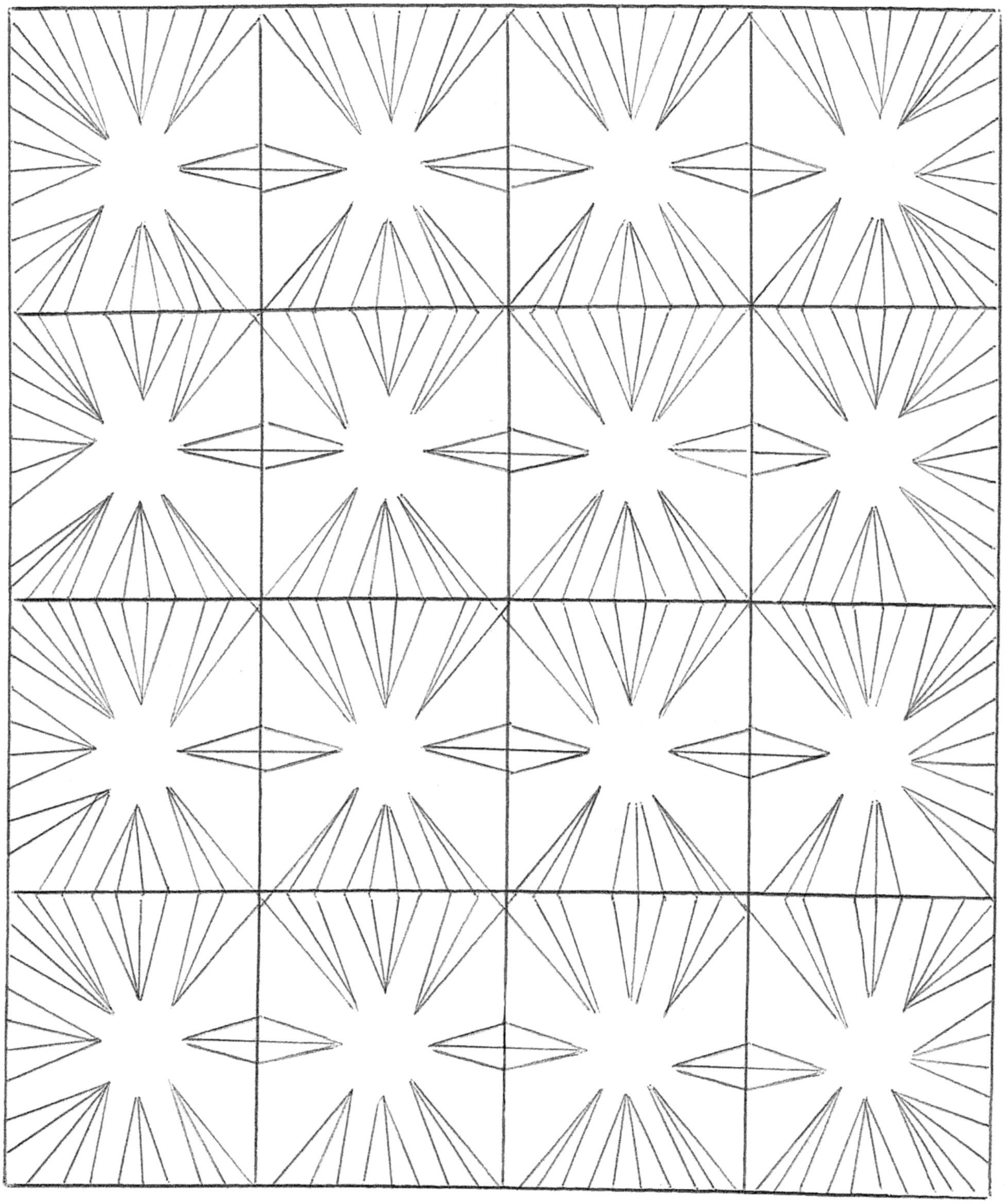

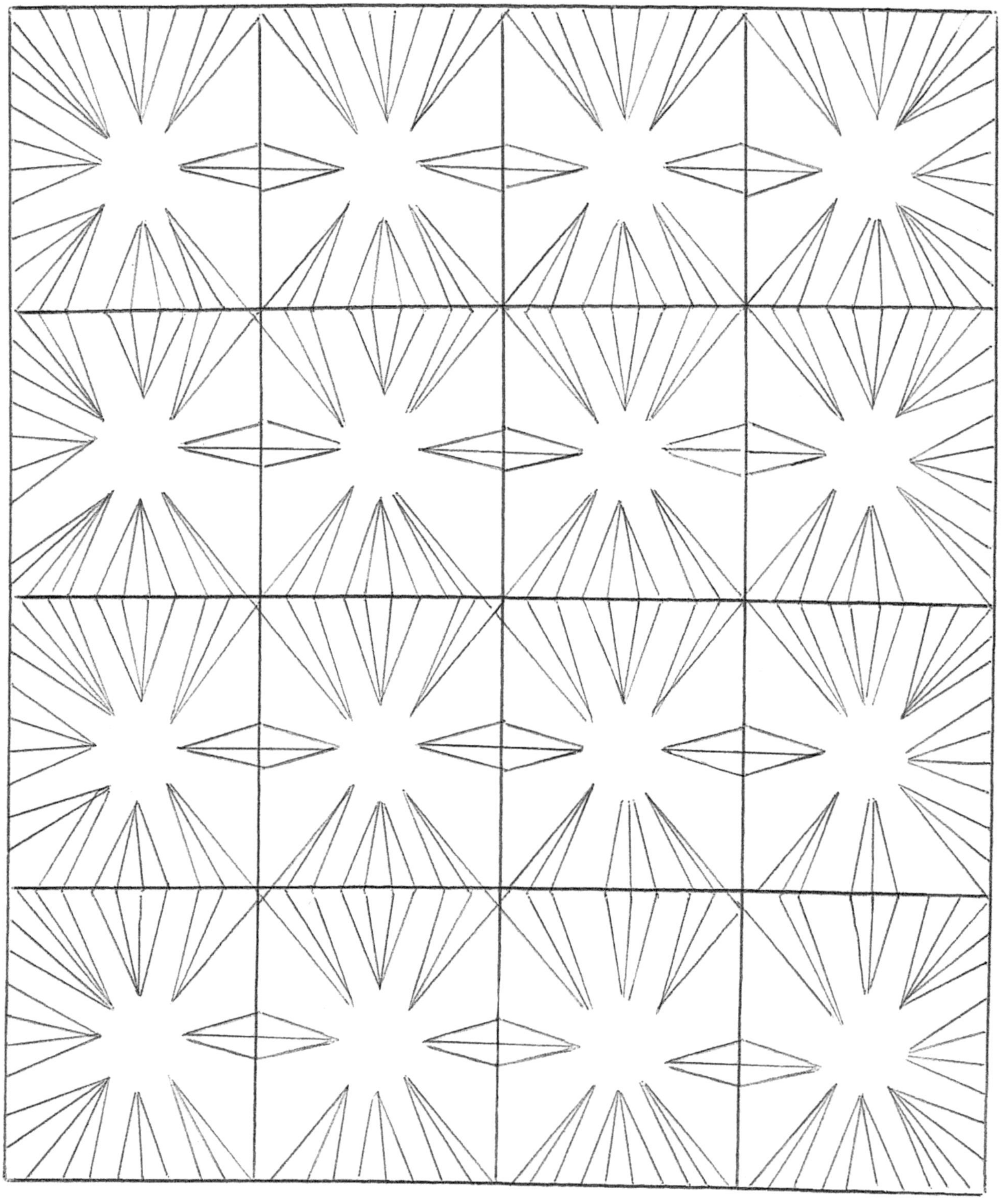

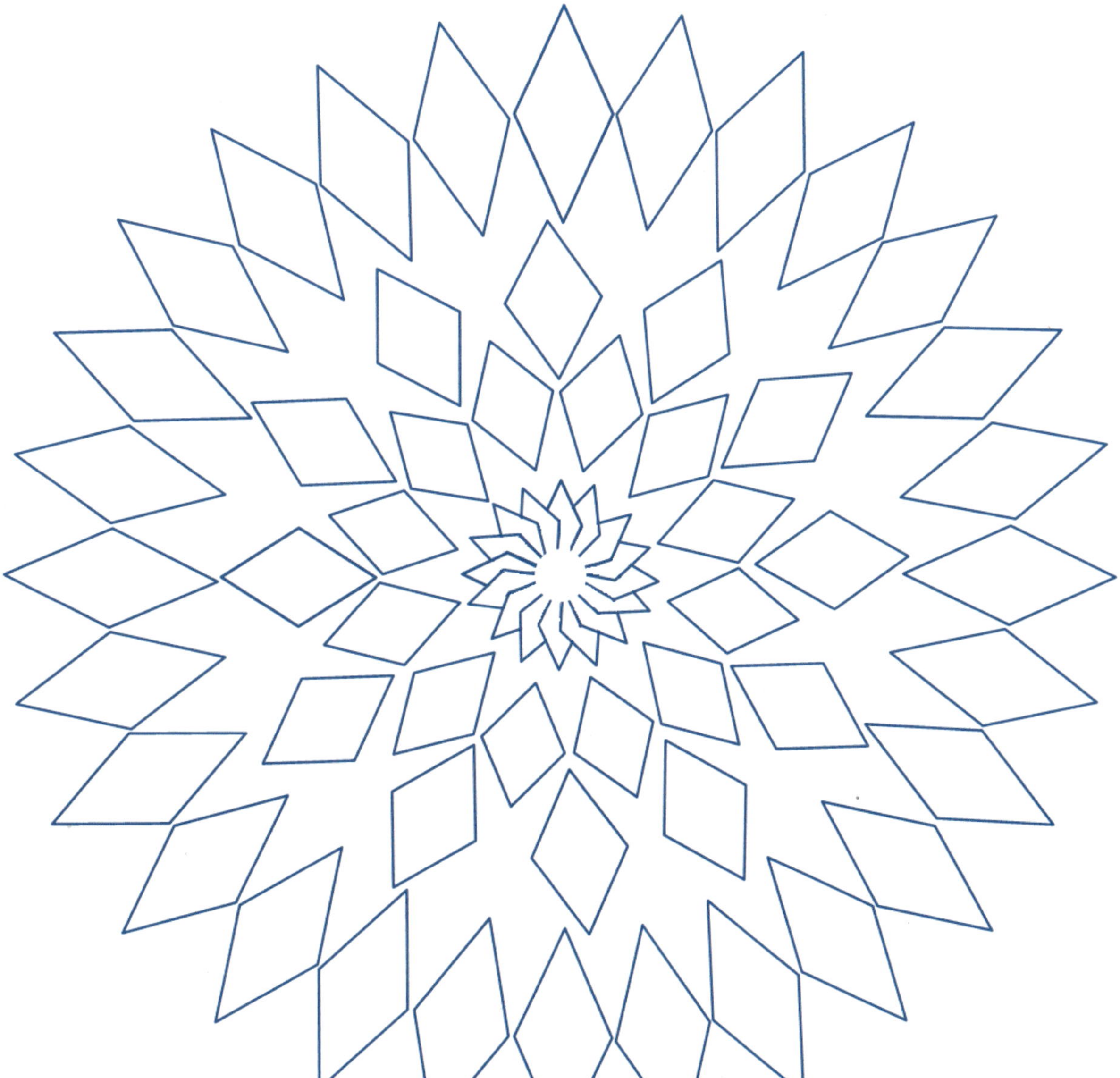

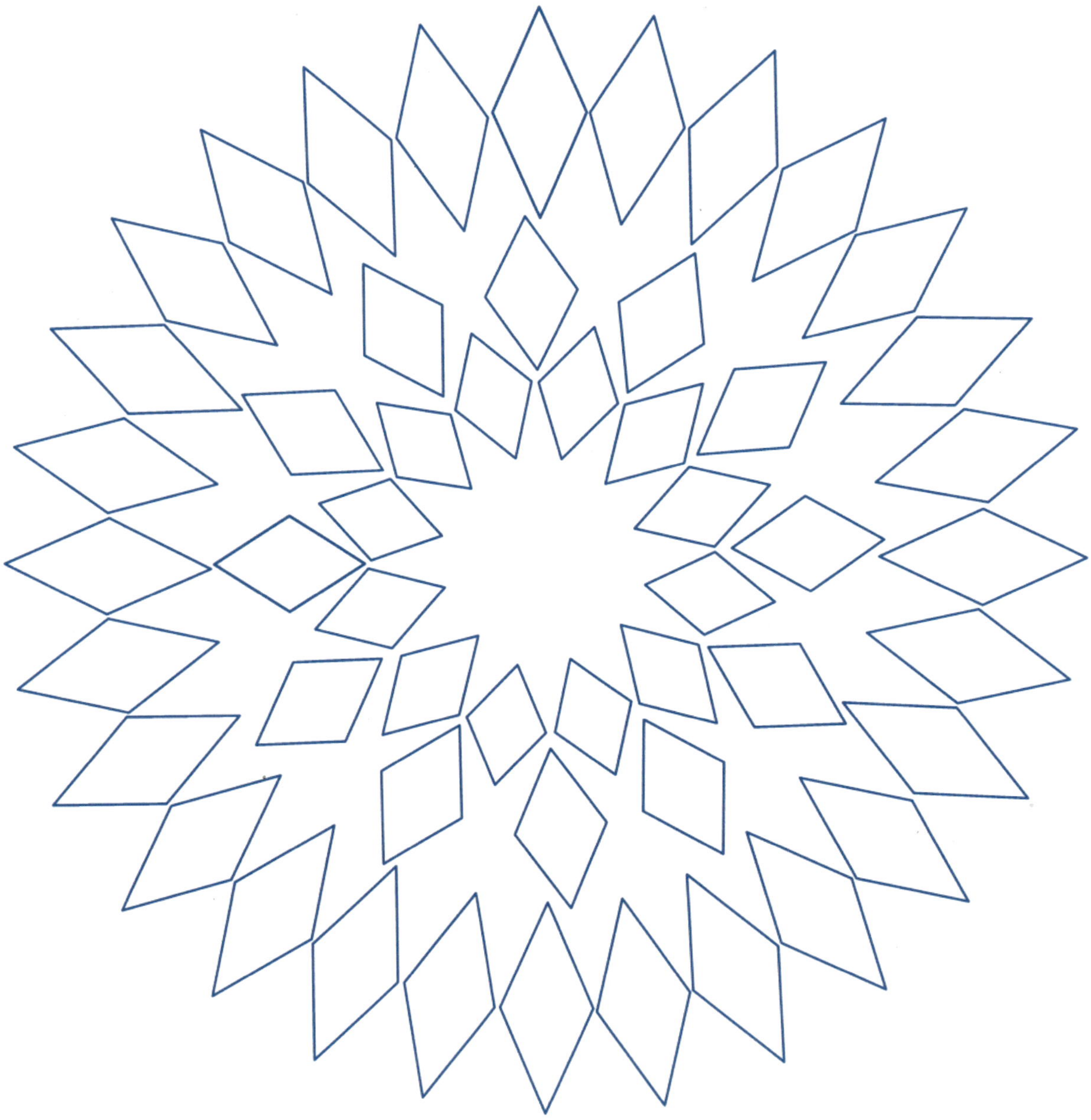

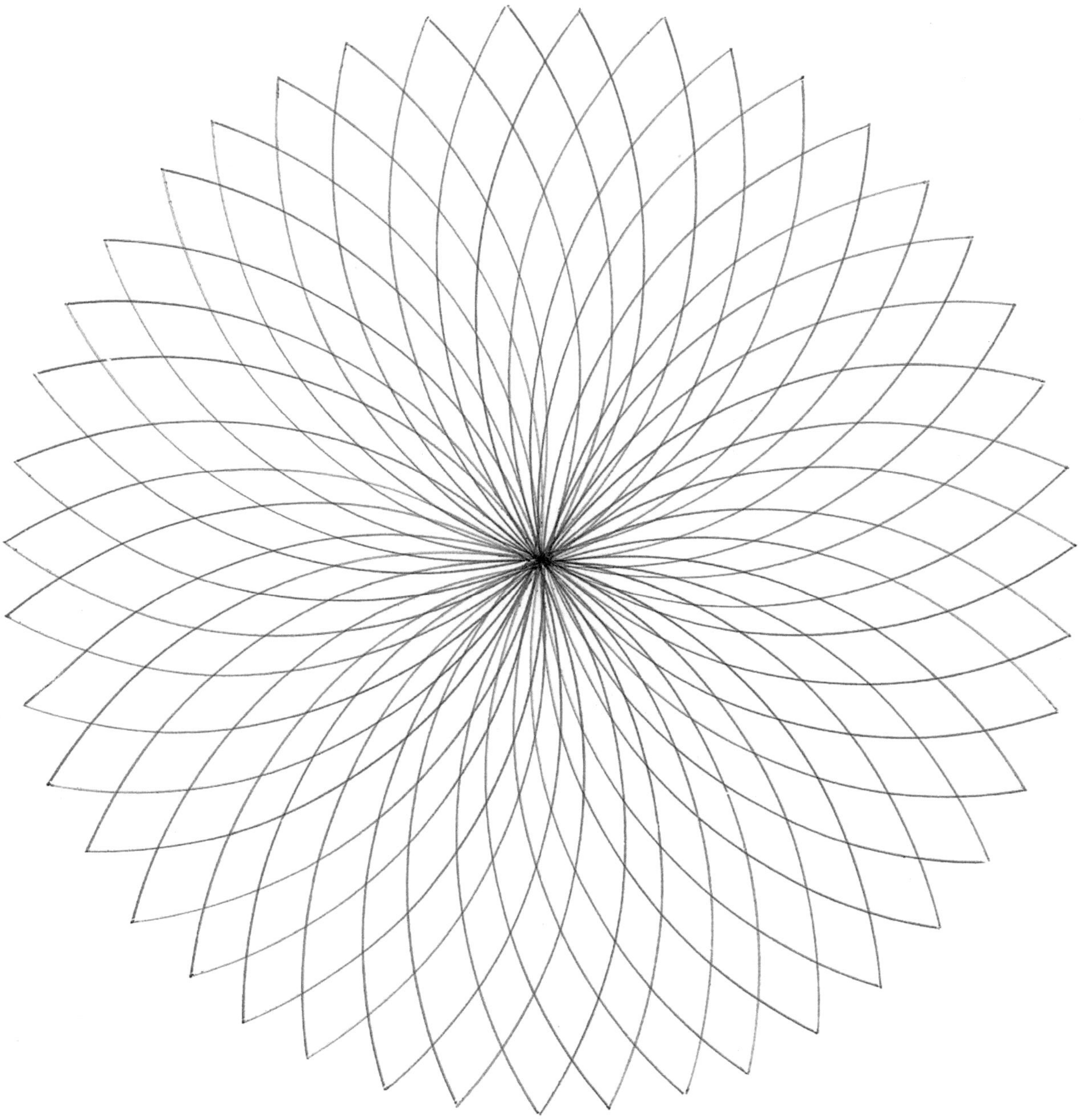

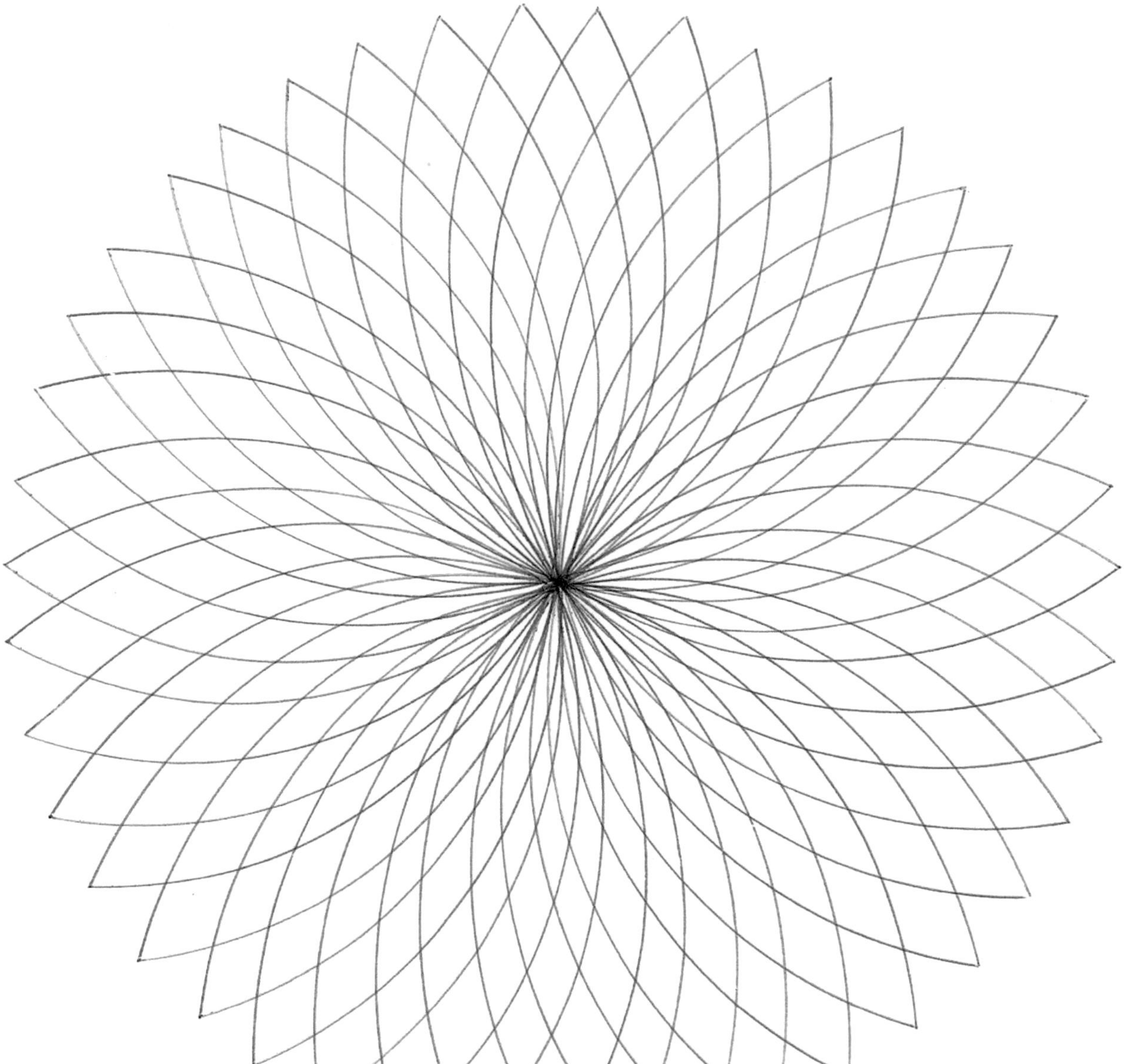

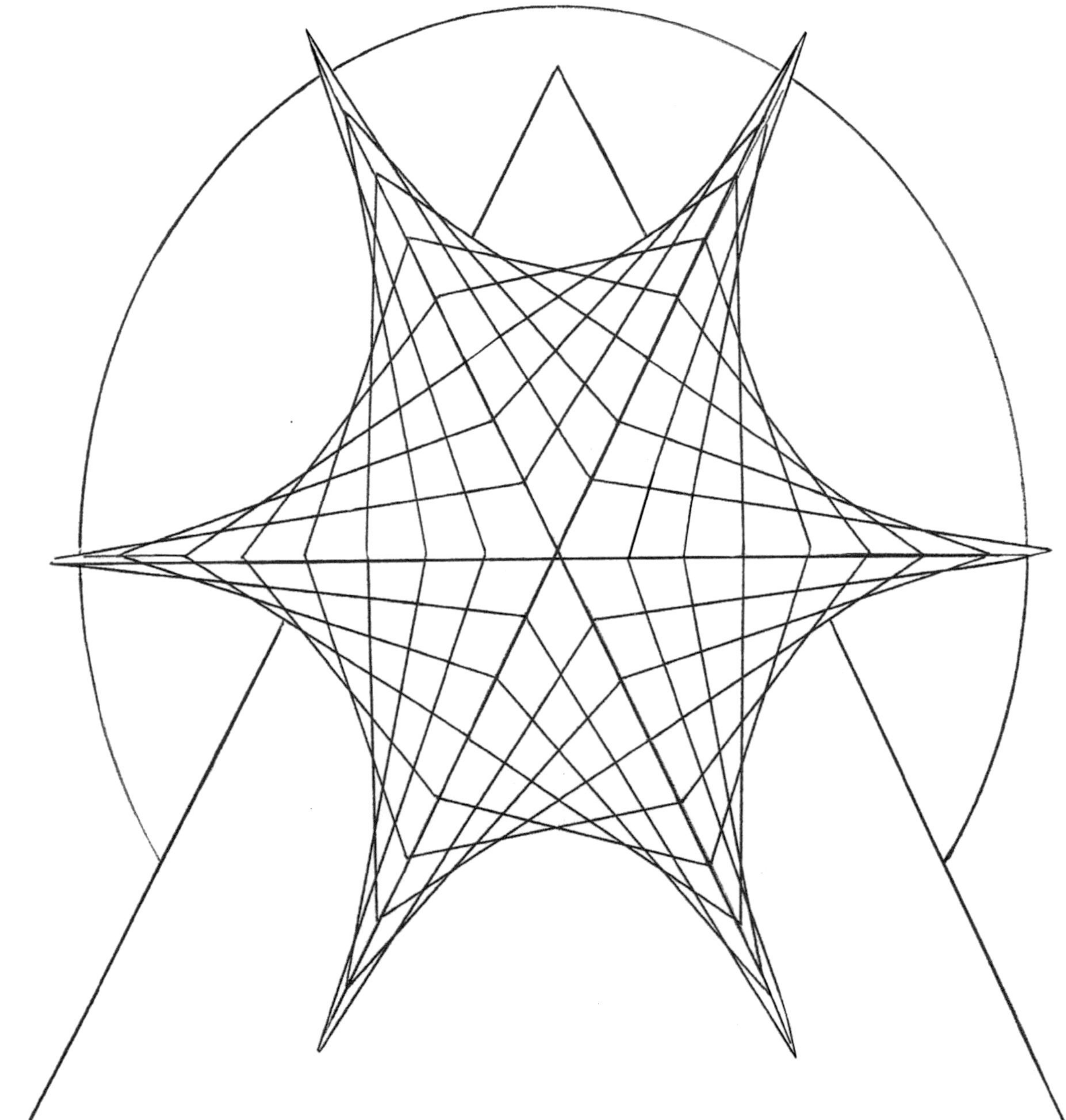

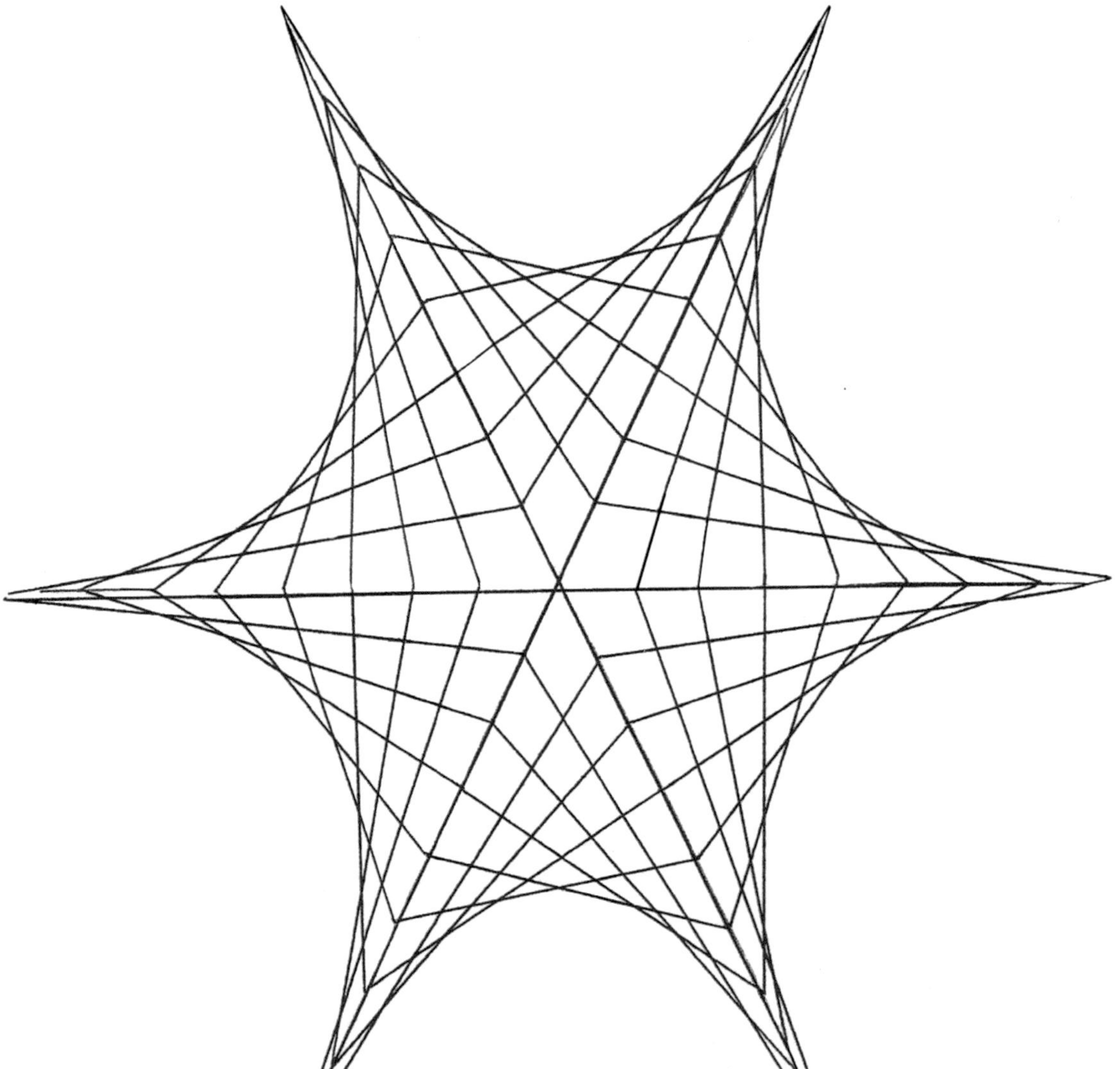

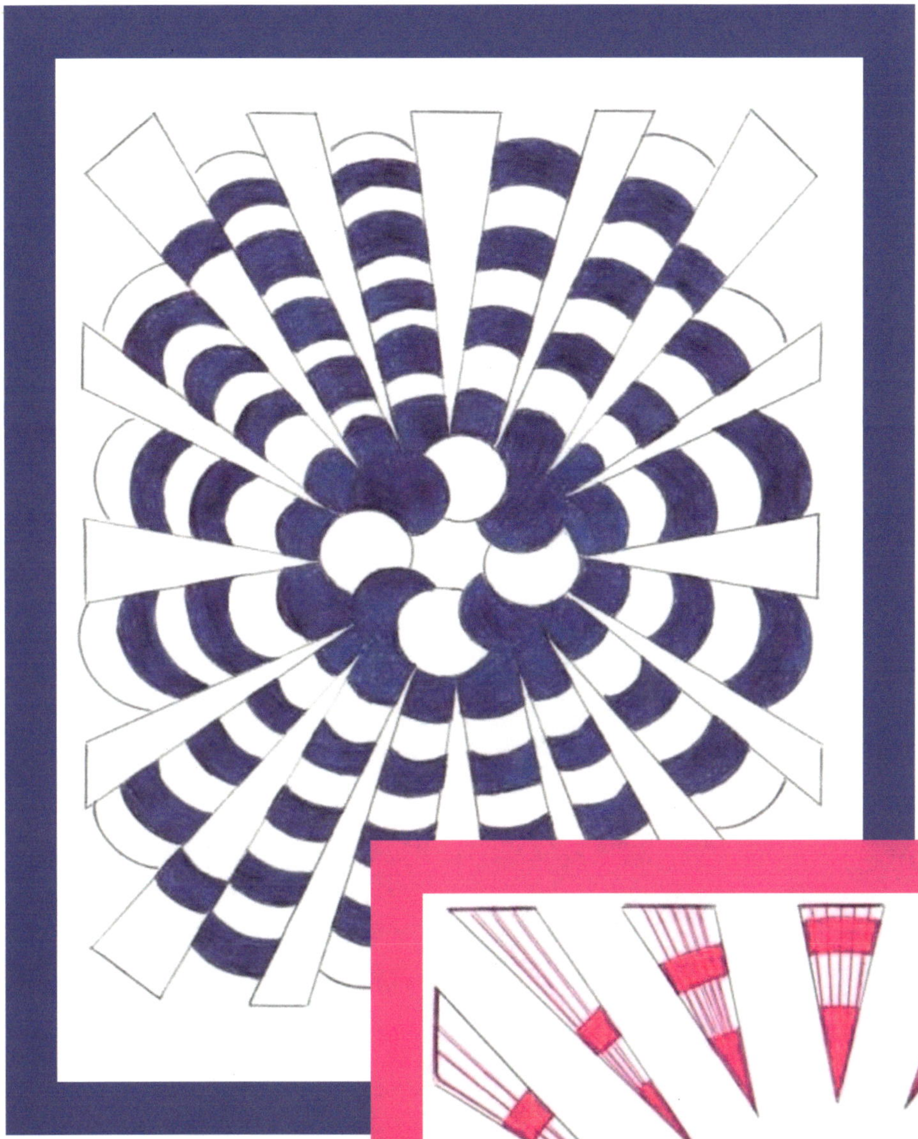

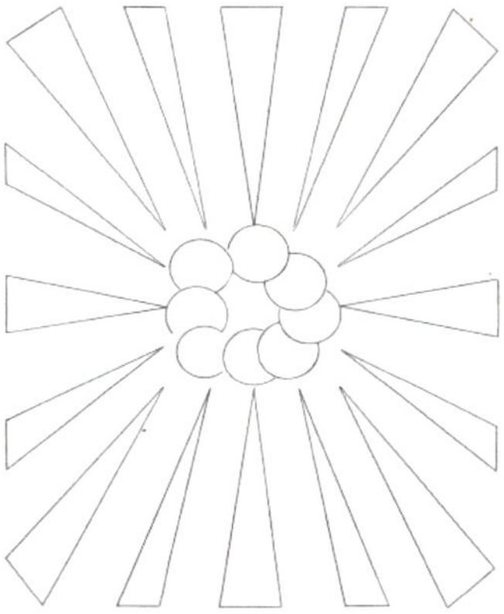

Purple Rhapsody

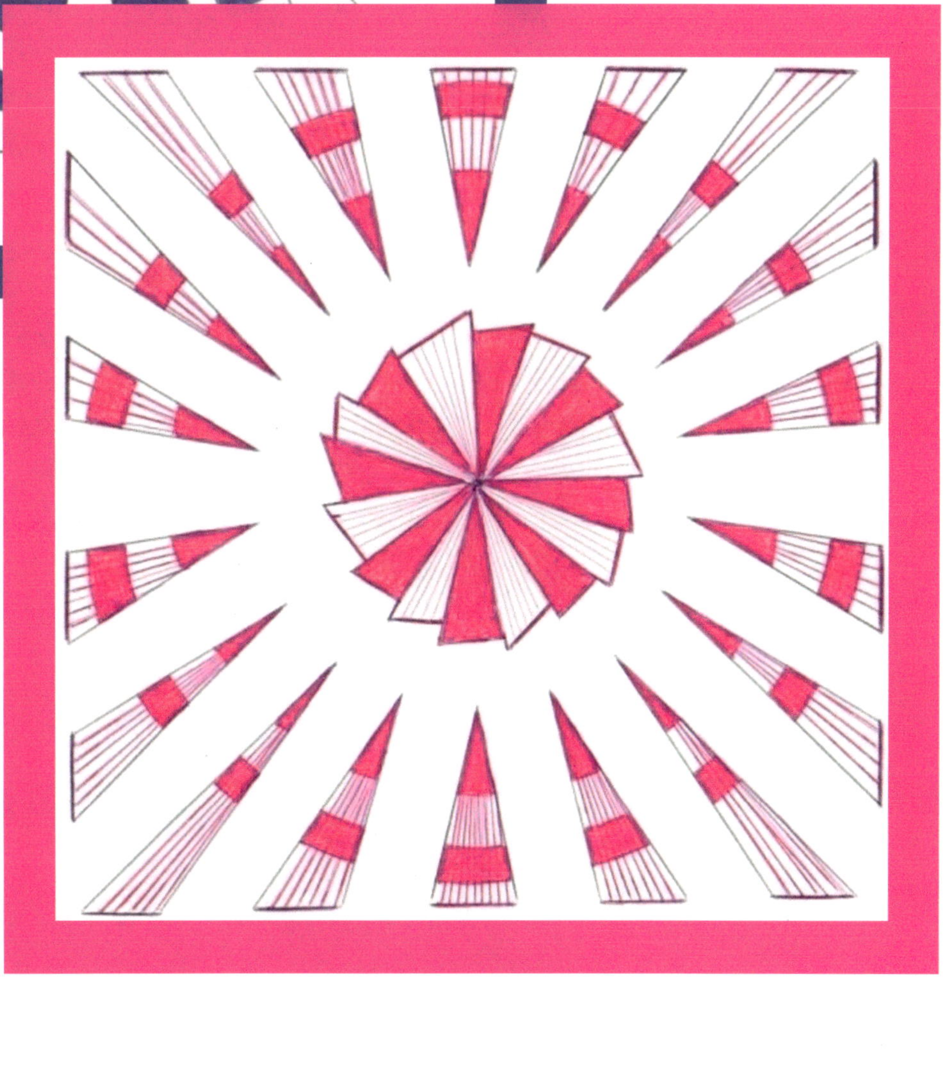

Fuchsia Sparks

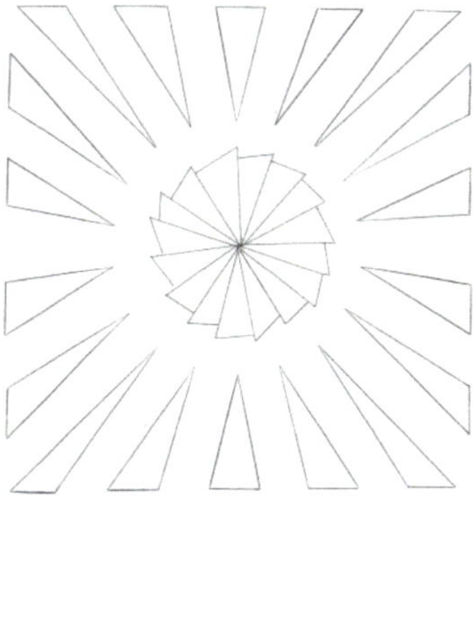

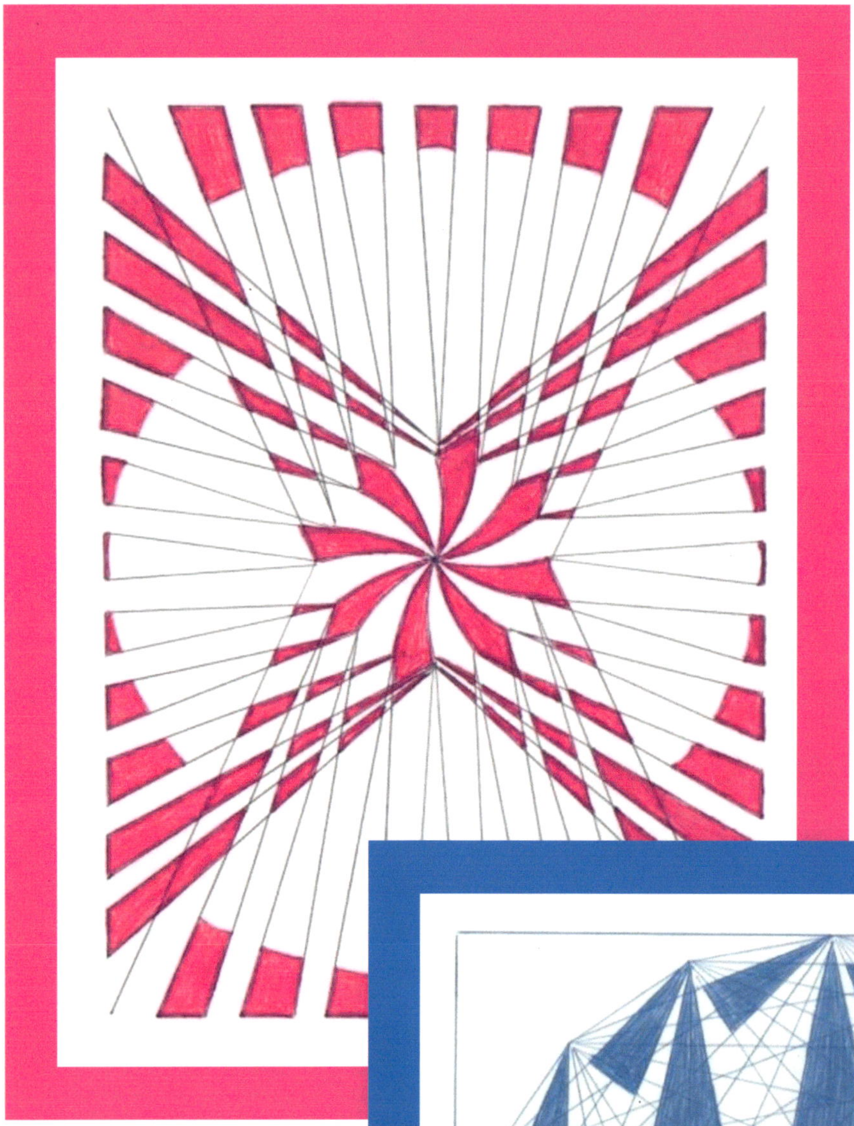

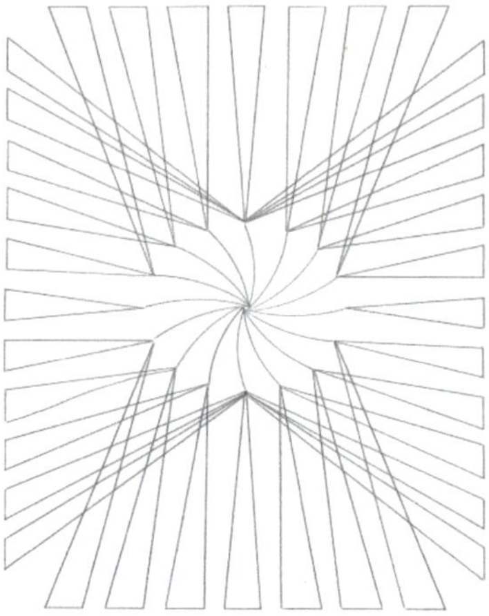

Sonata

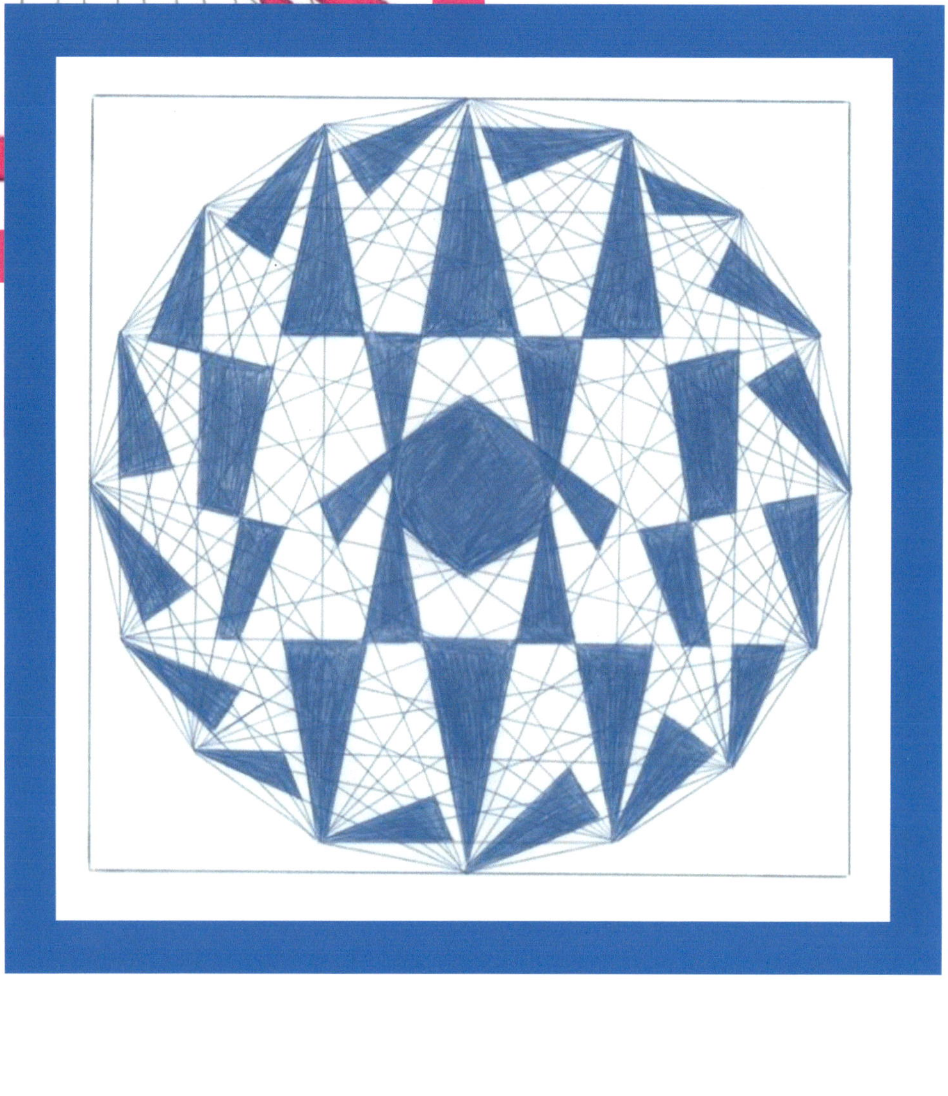

A Wild Ride

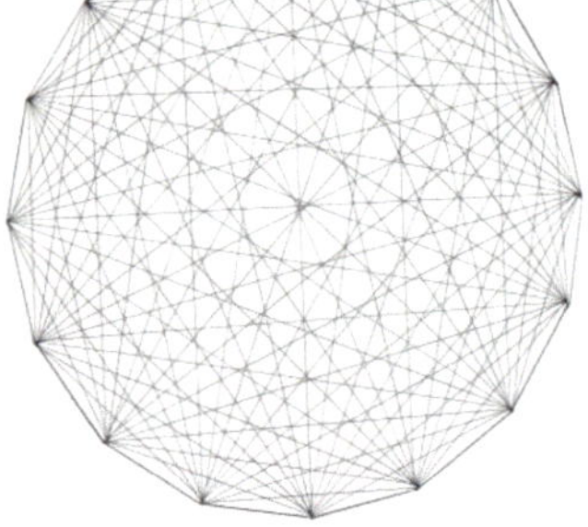

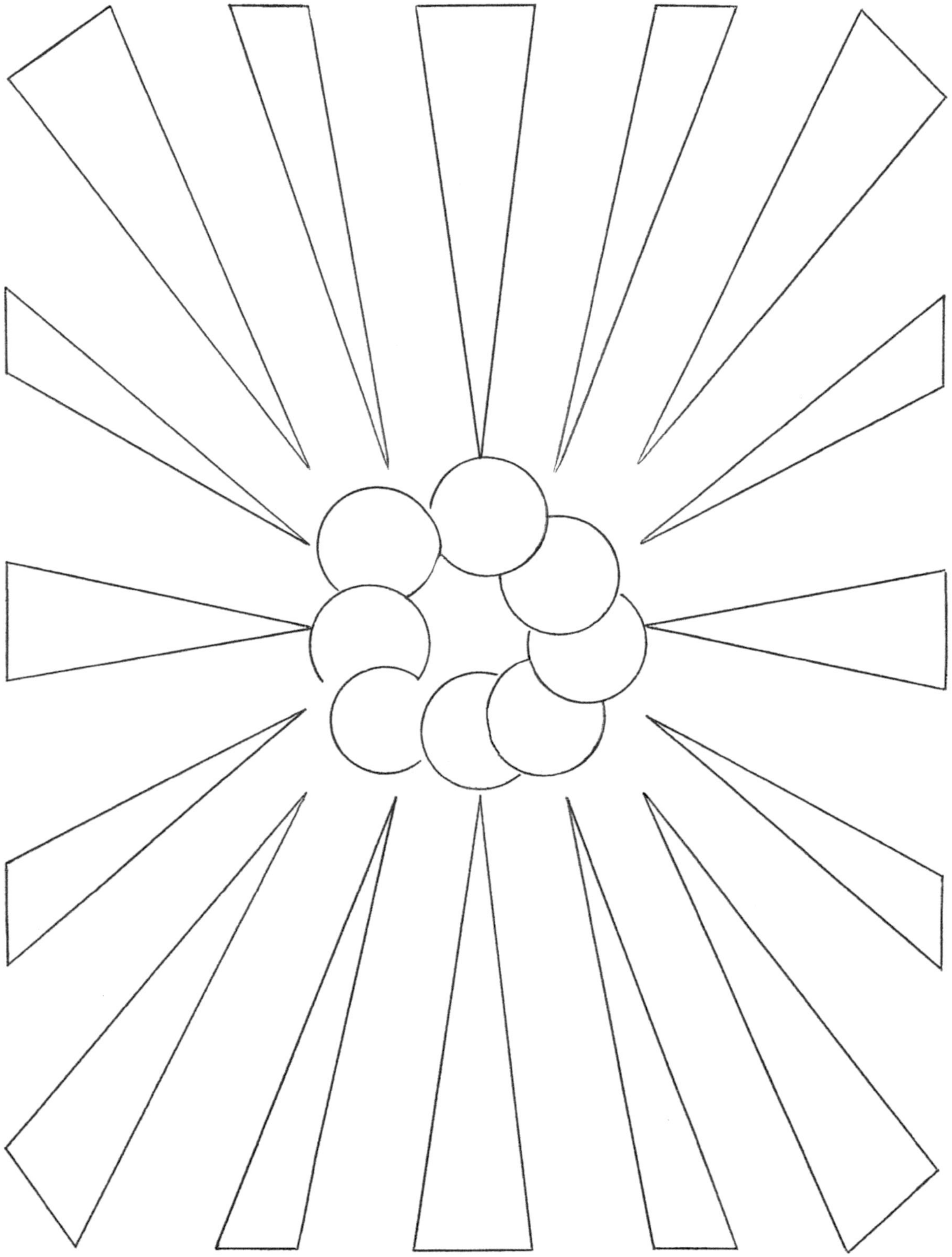

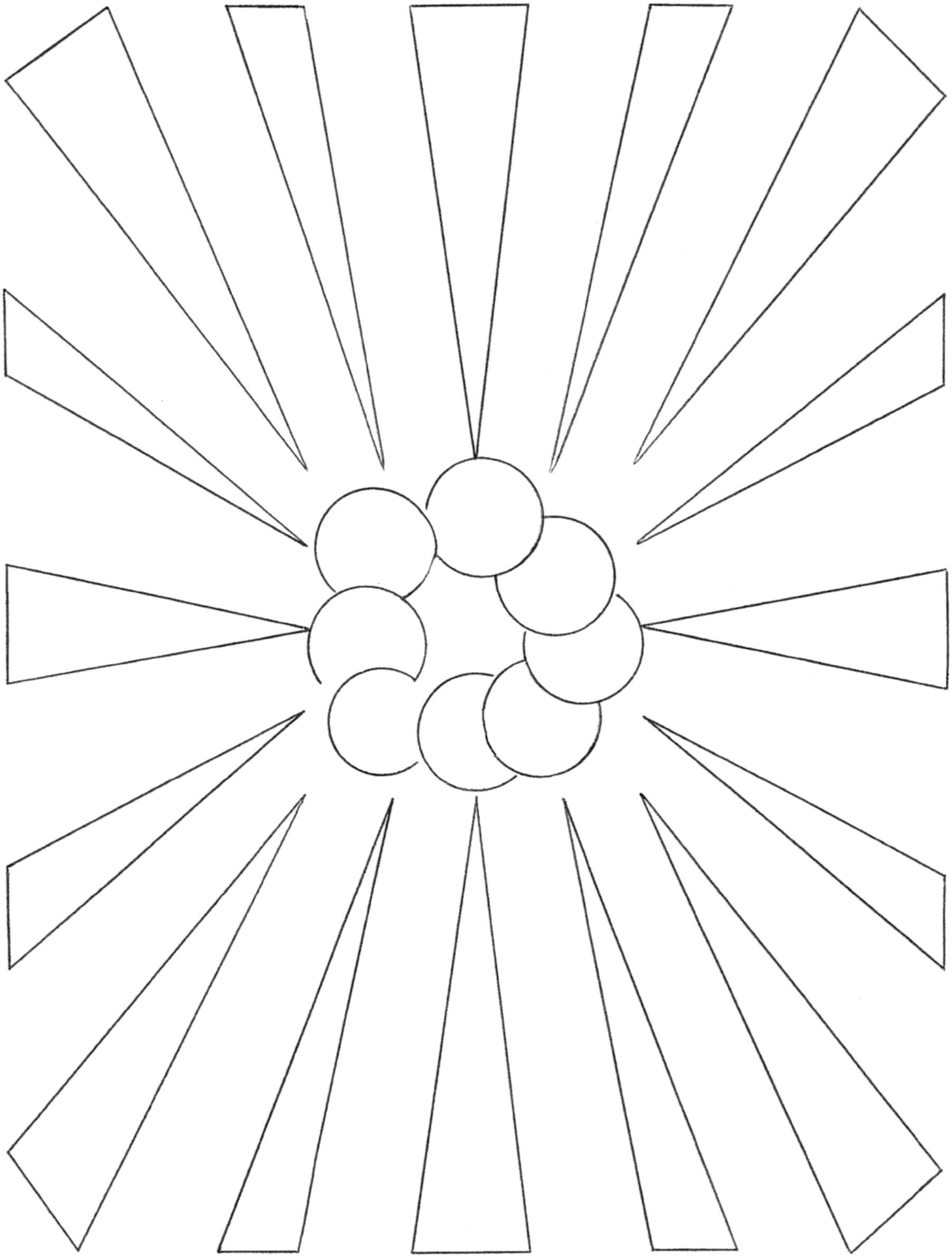

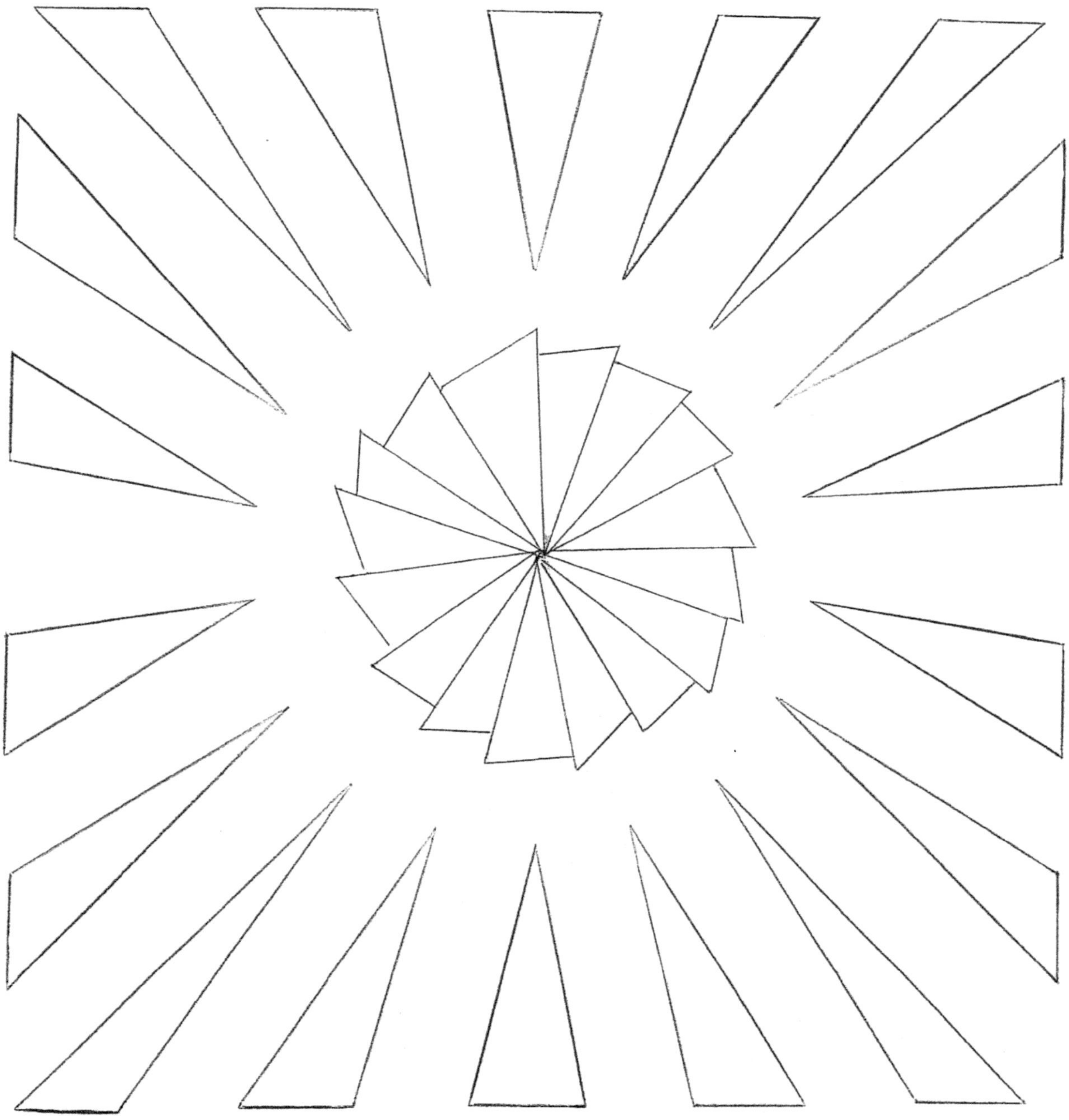

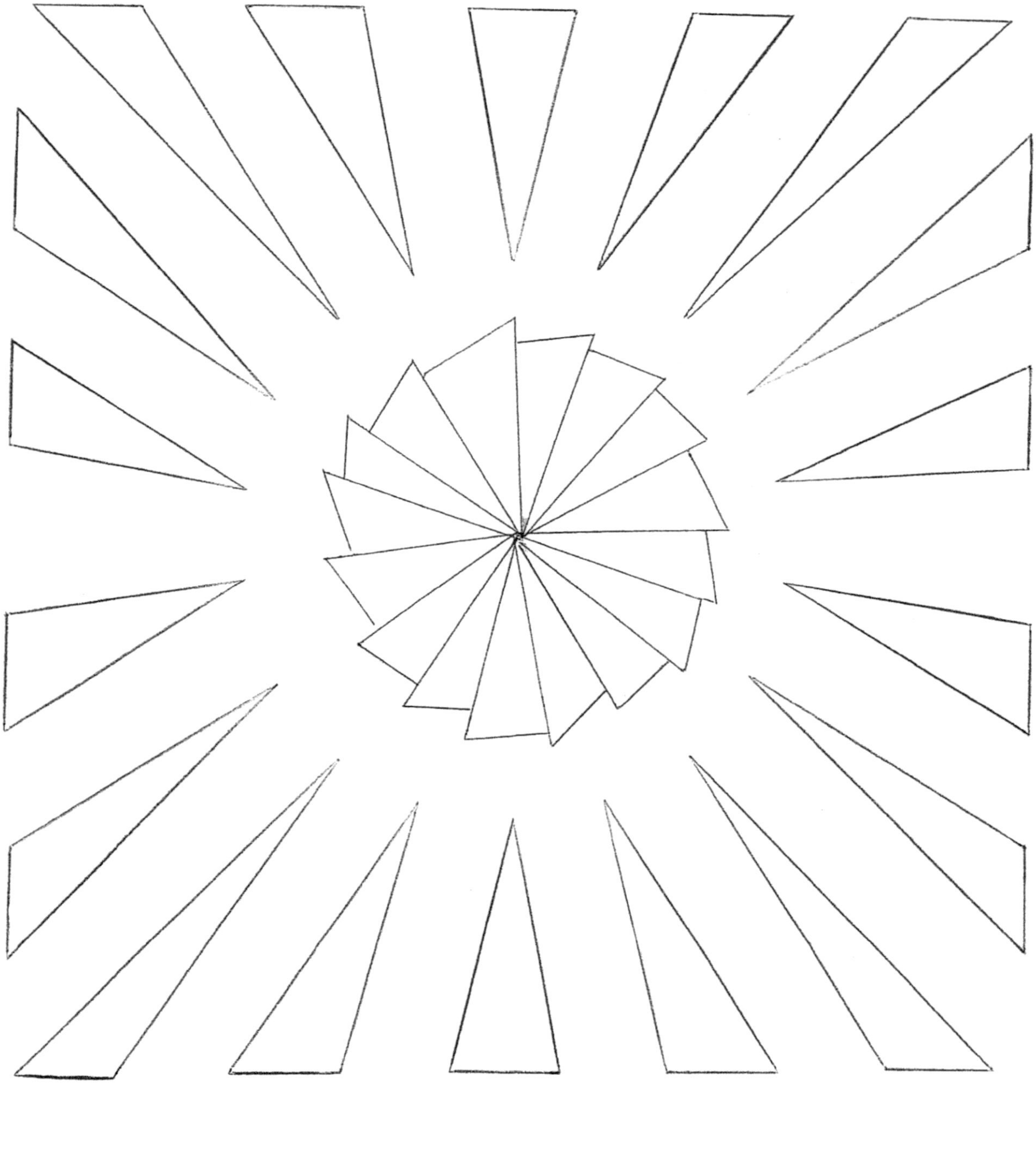

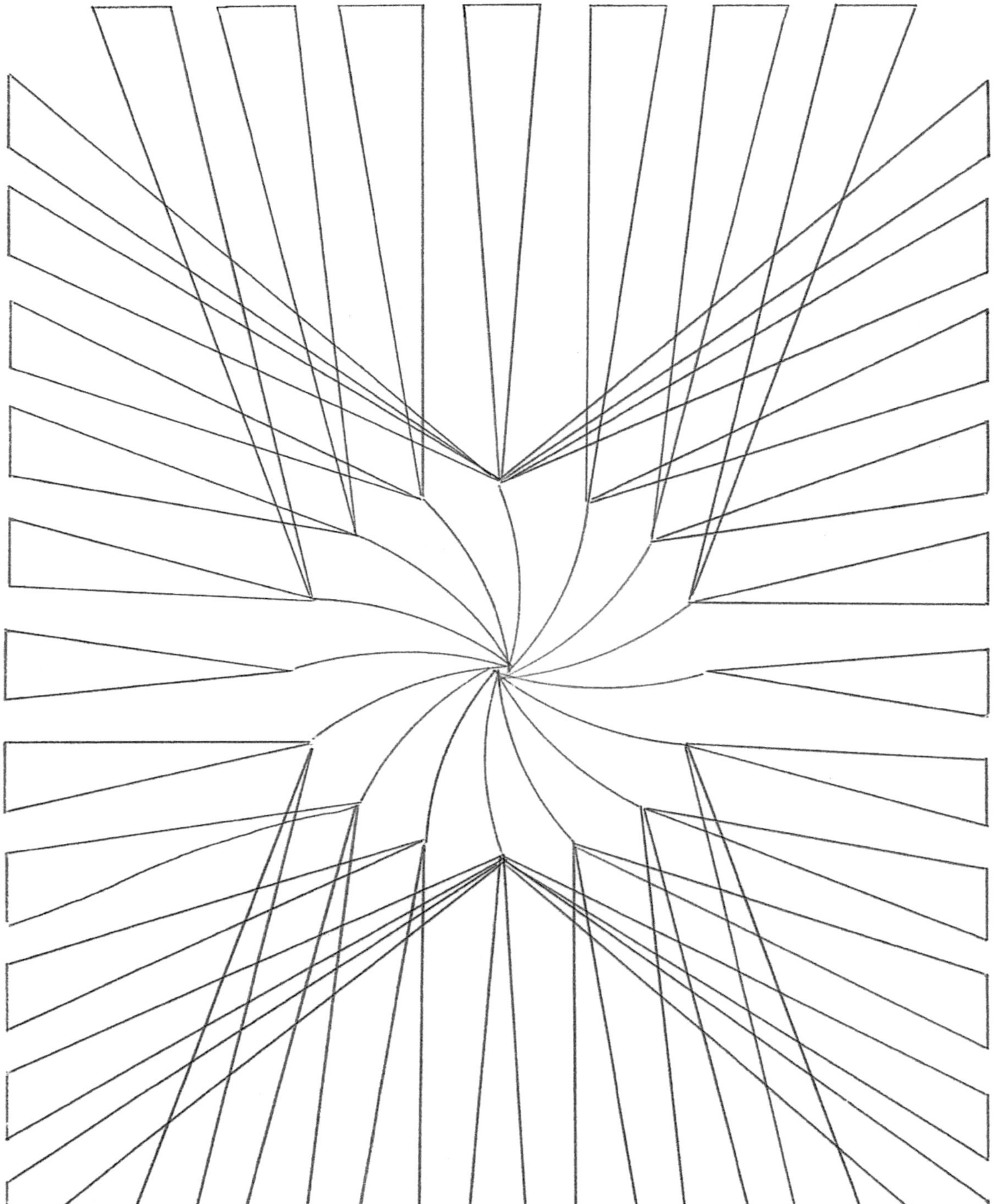

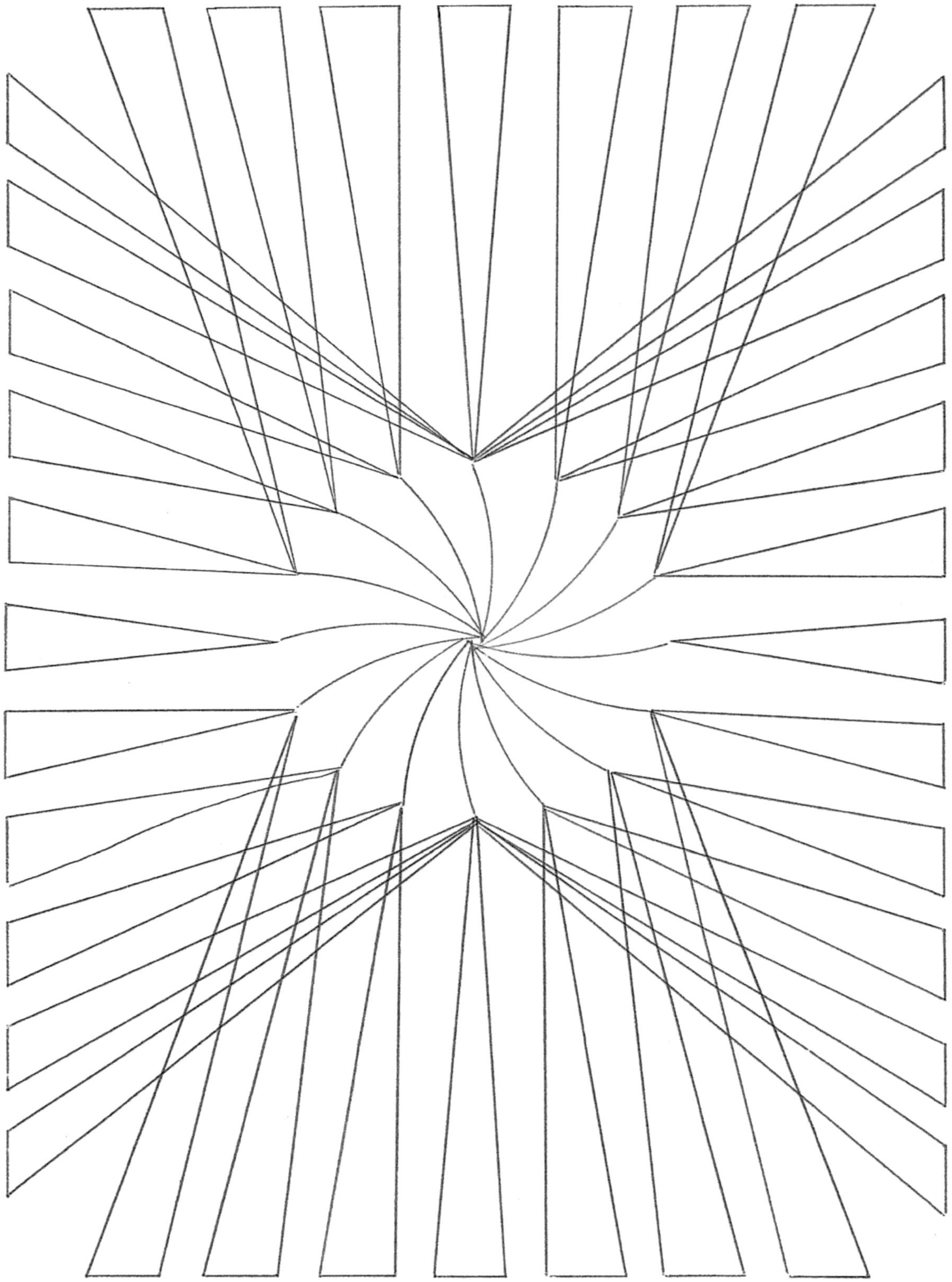

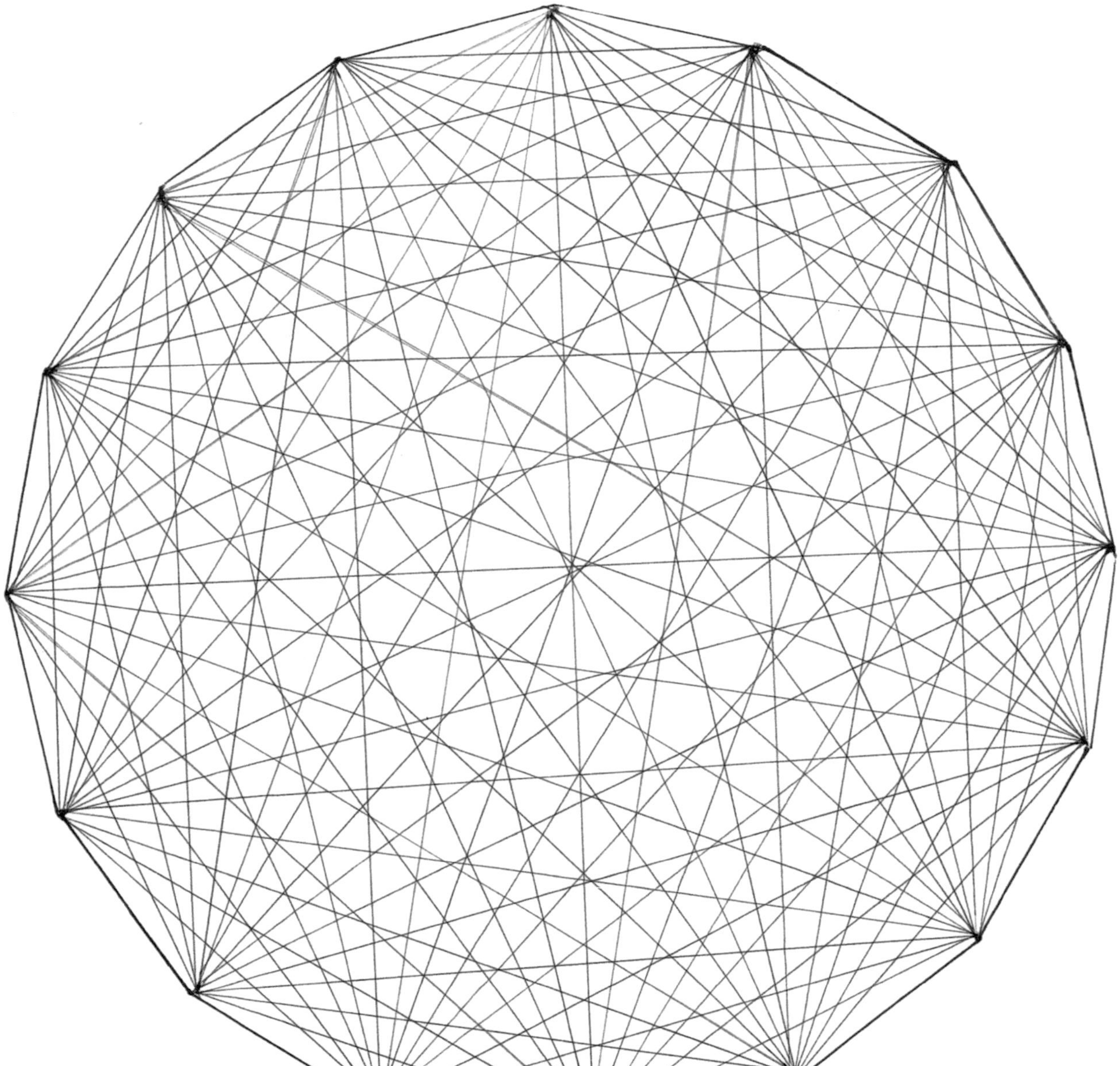

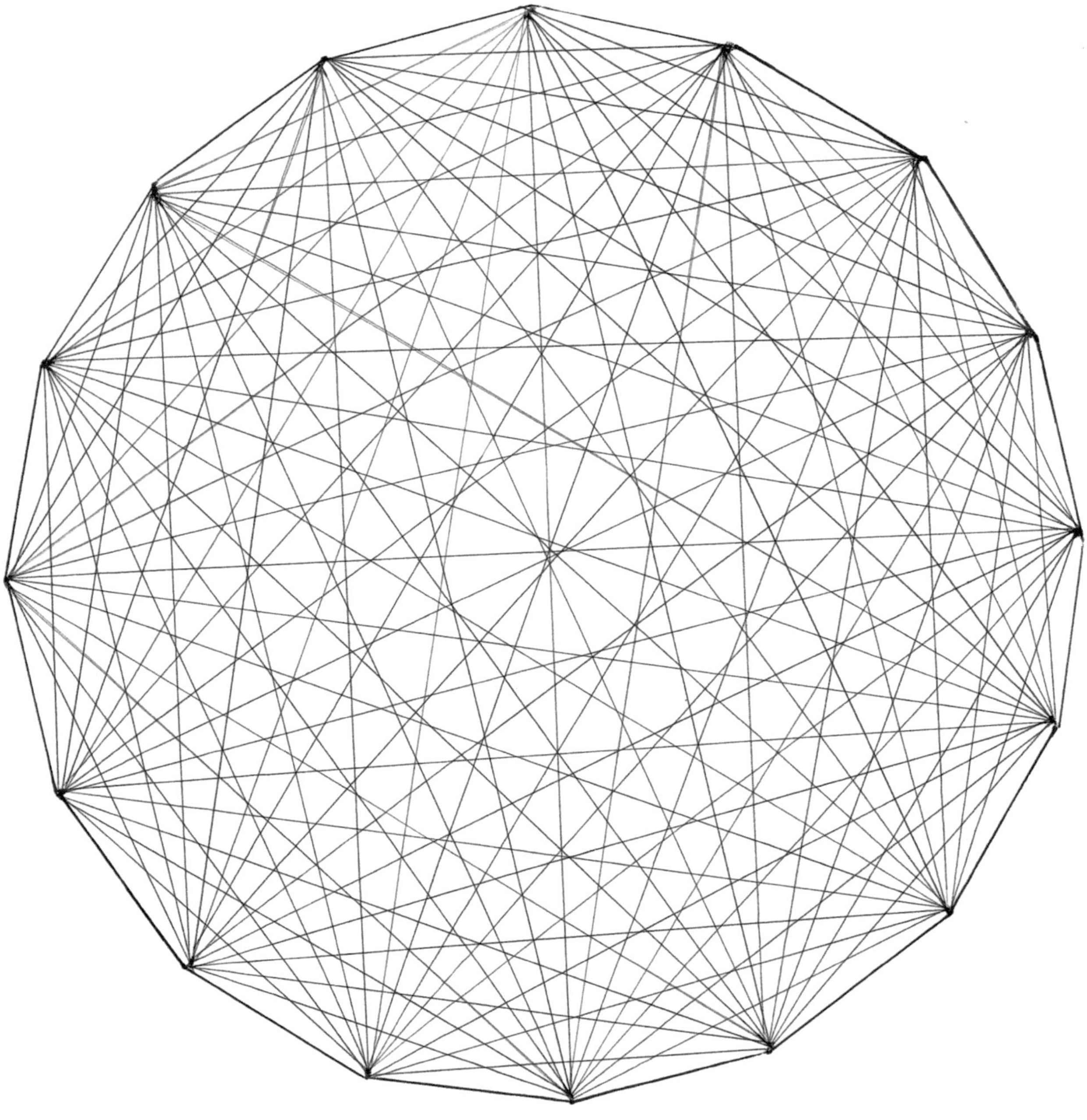

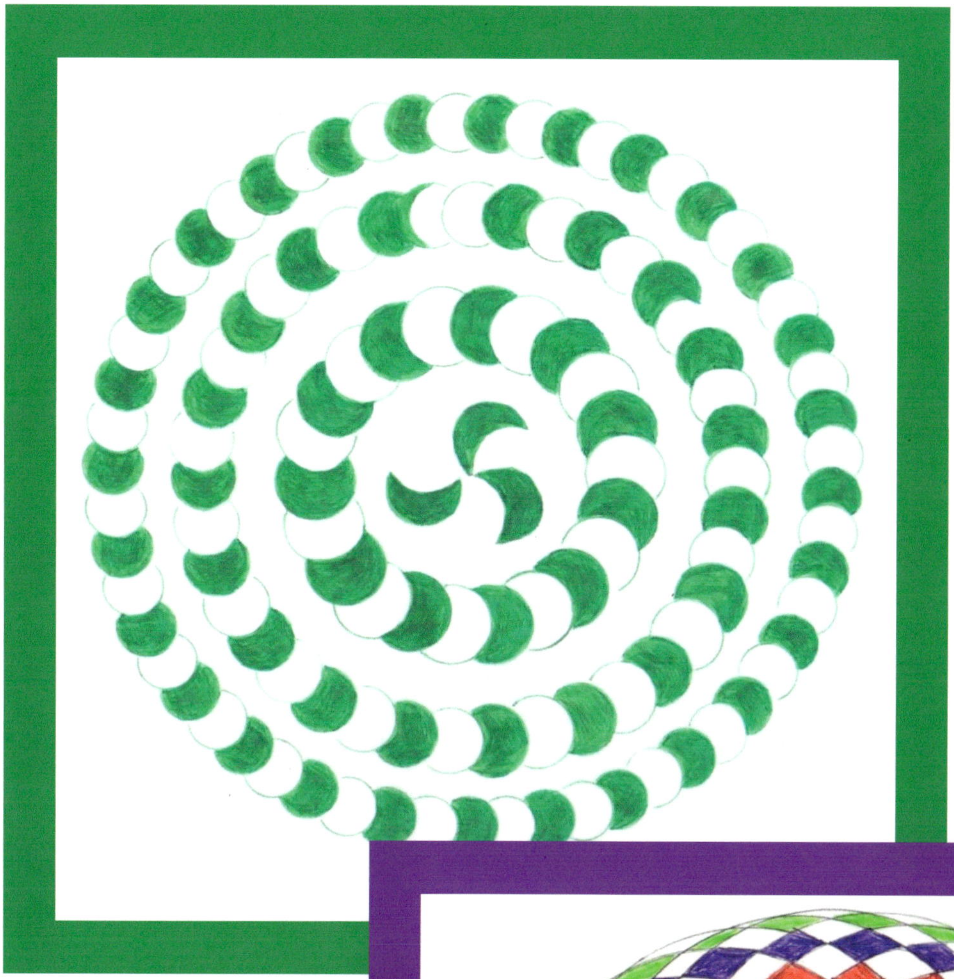

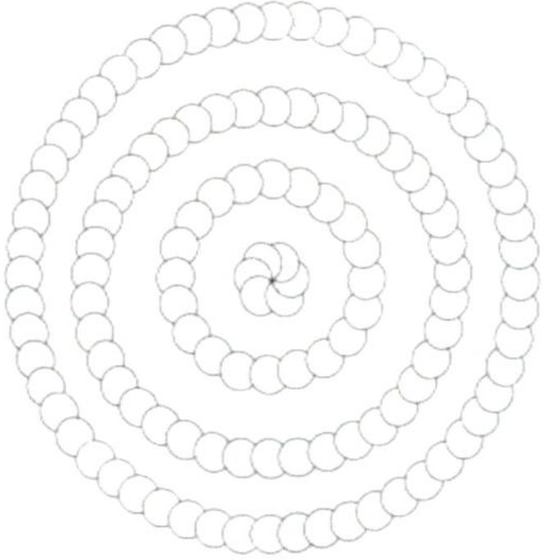

Circles Go Round

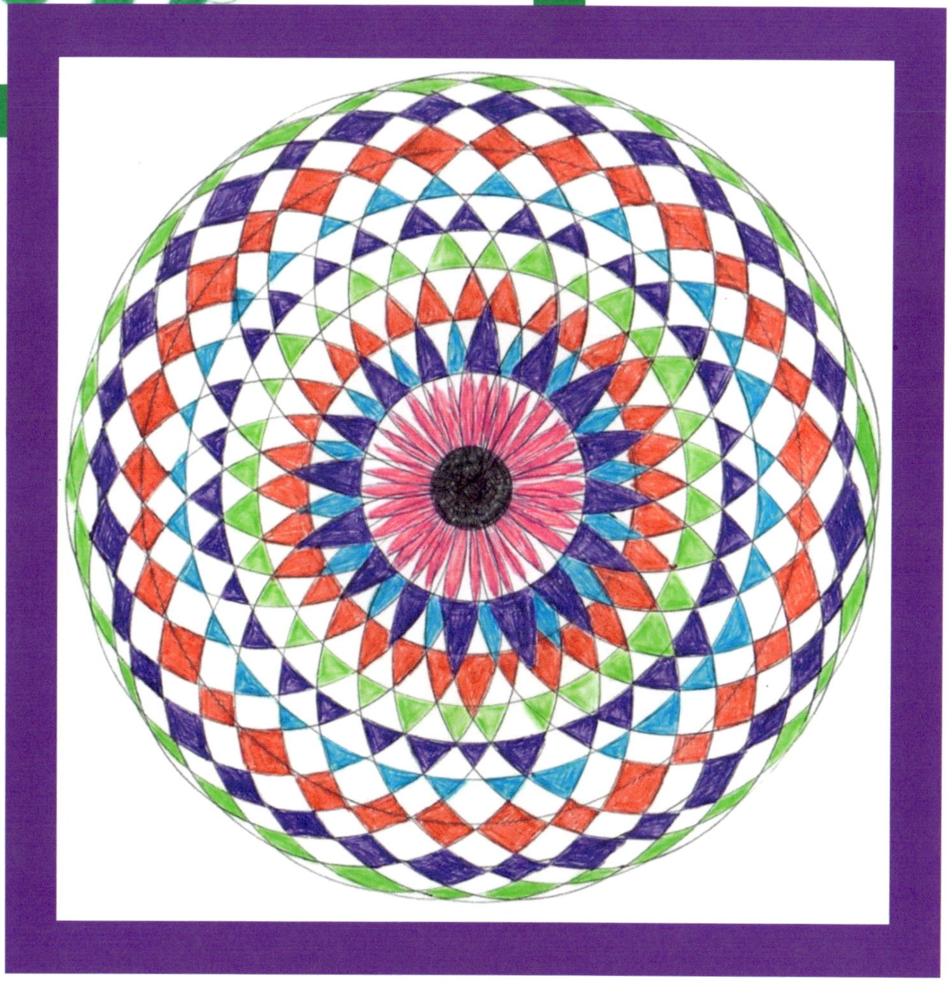

Fascination

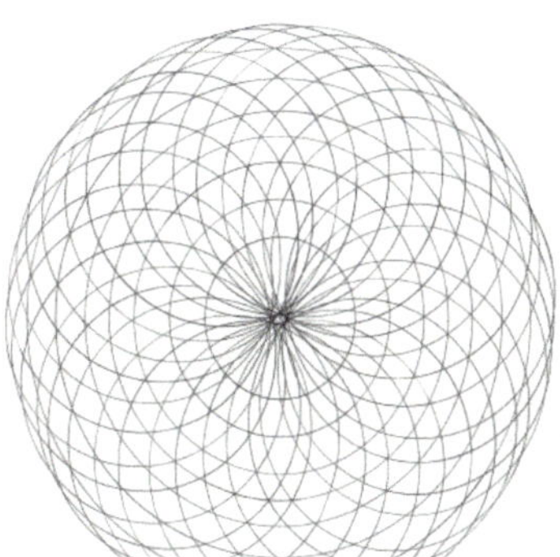

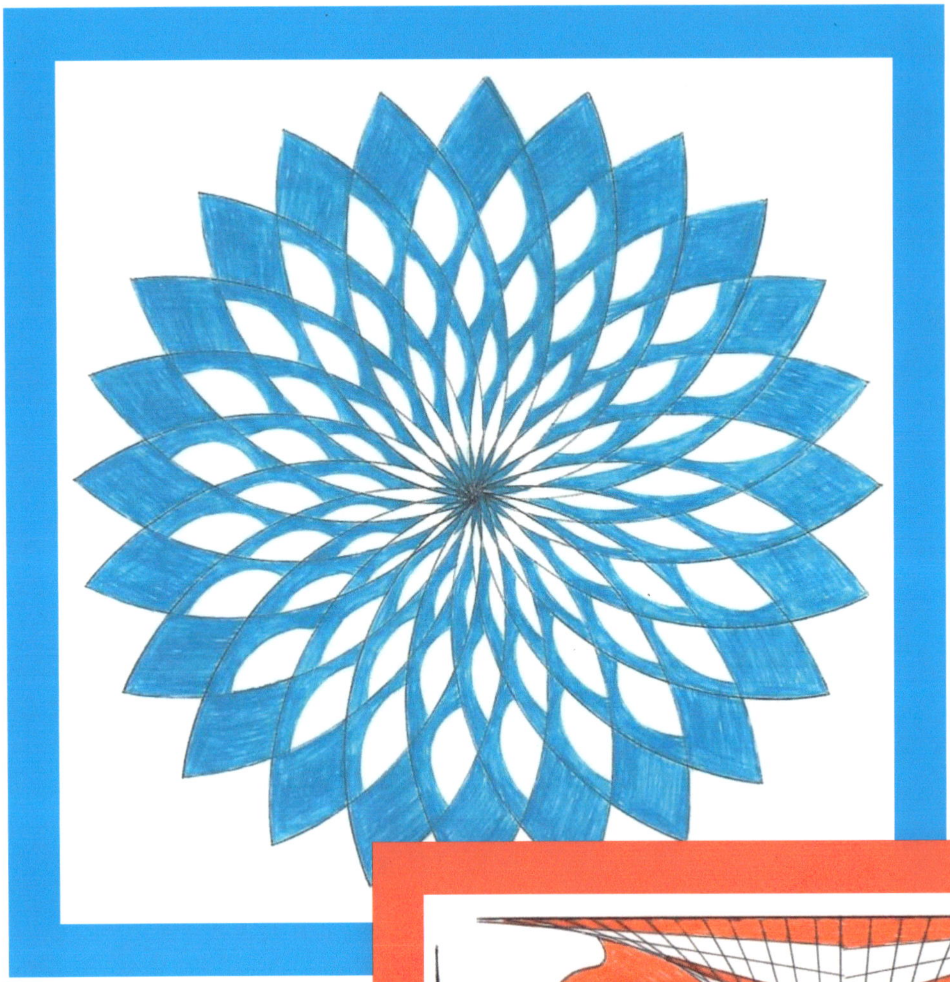

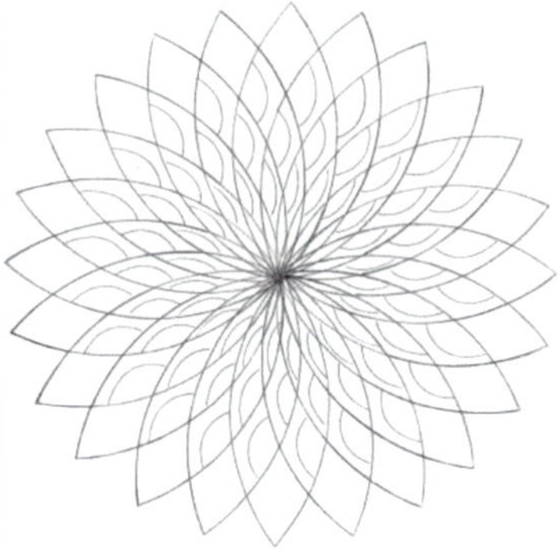

Blue Lace

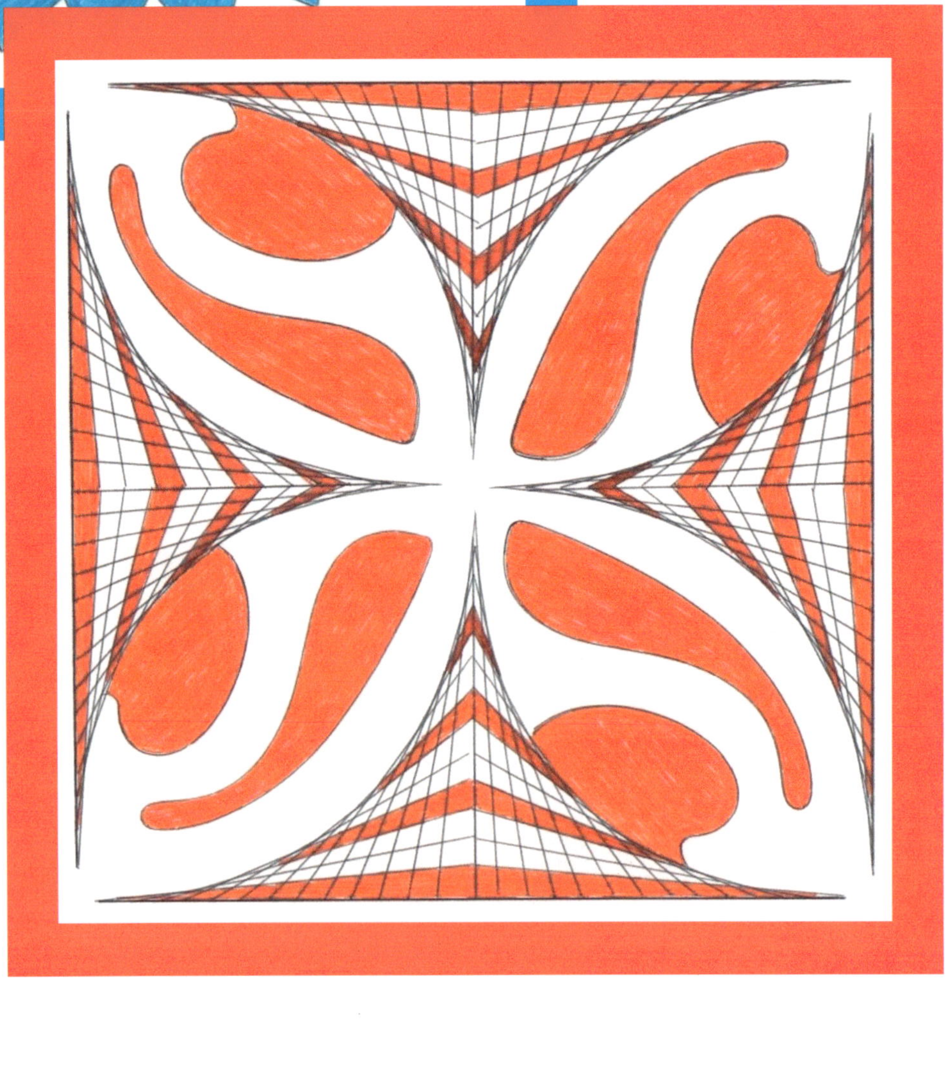

Orange Teardrops

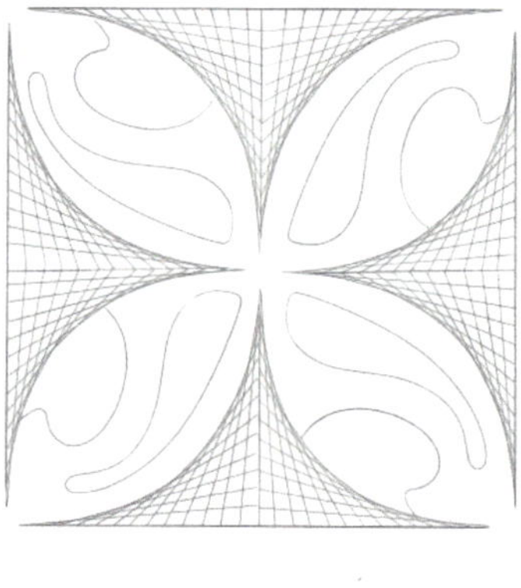

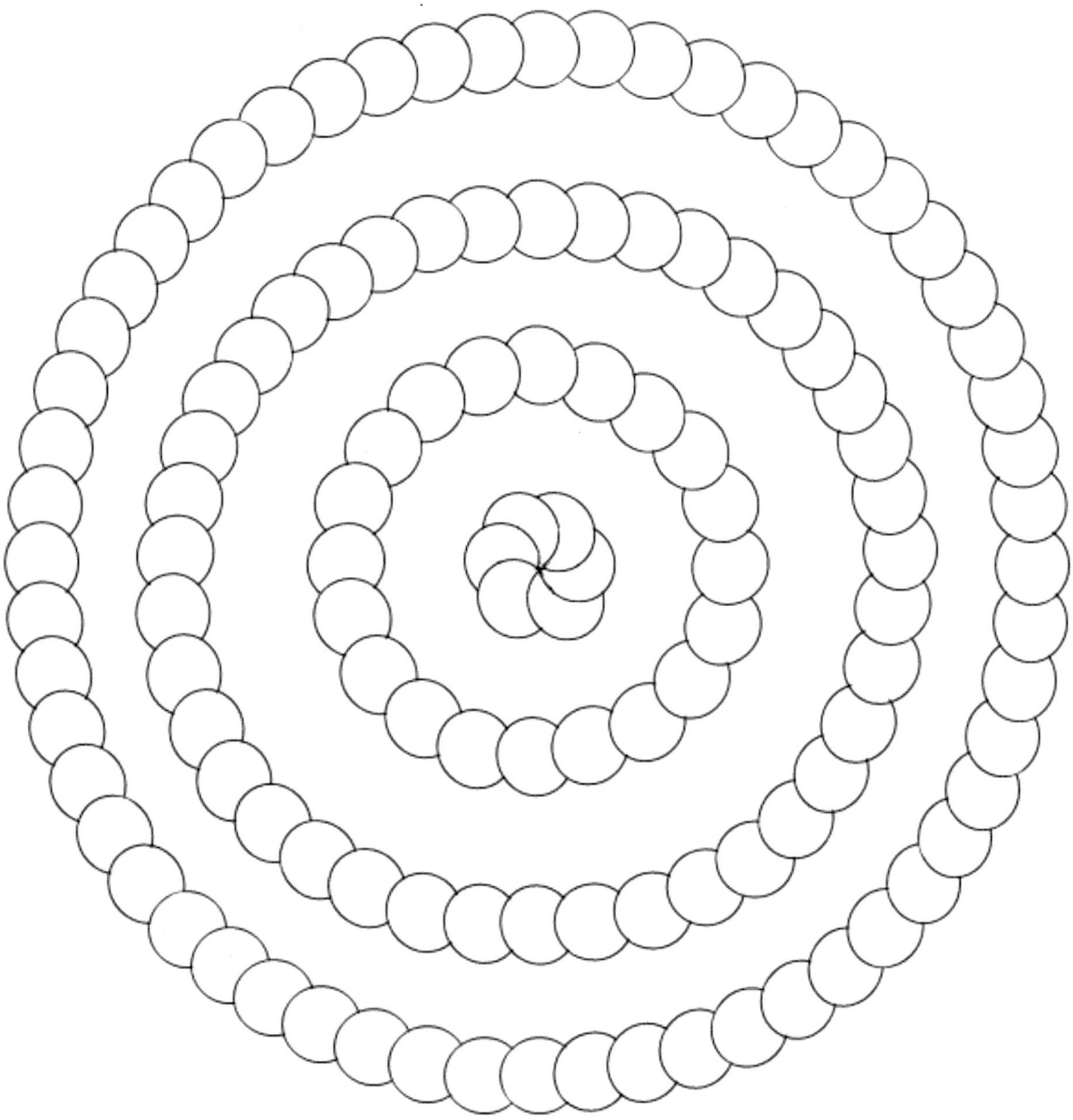

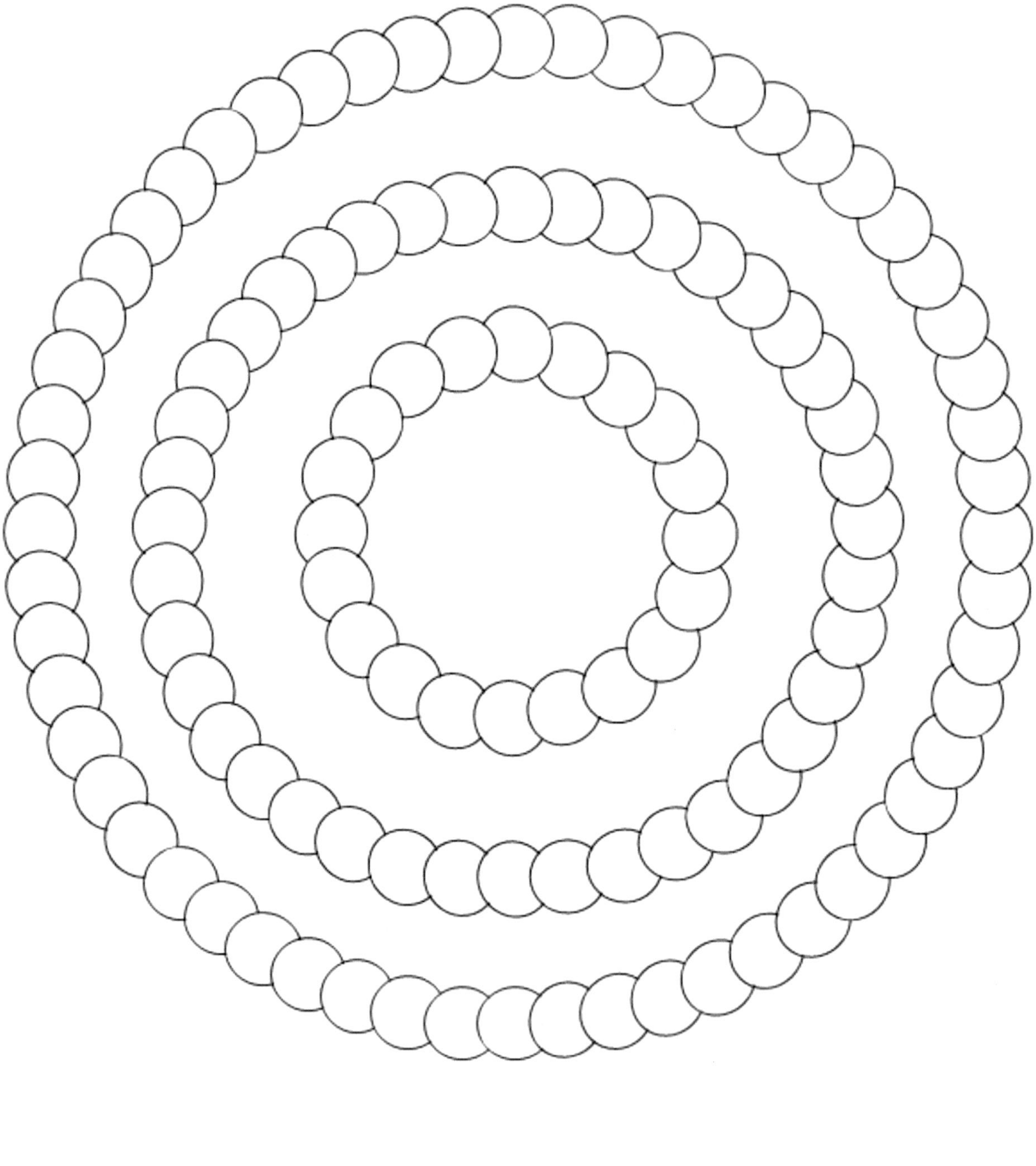

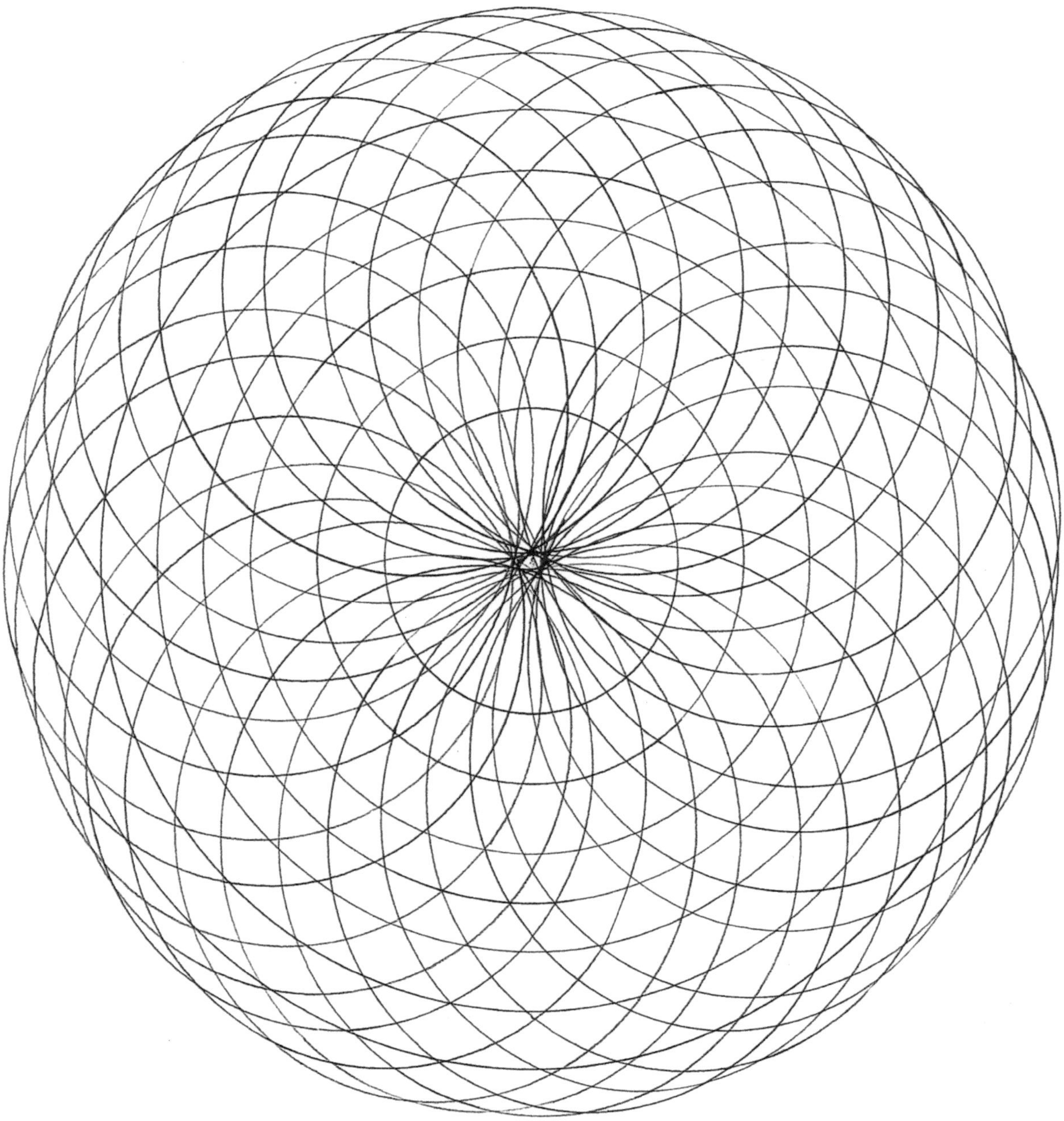

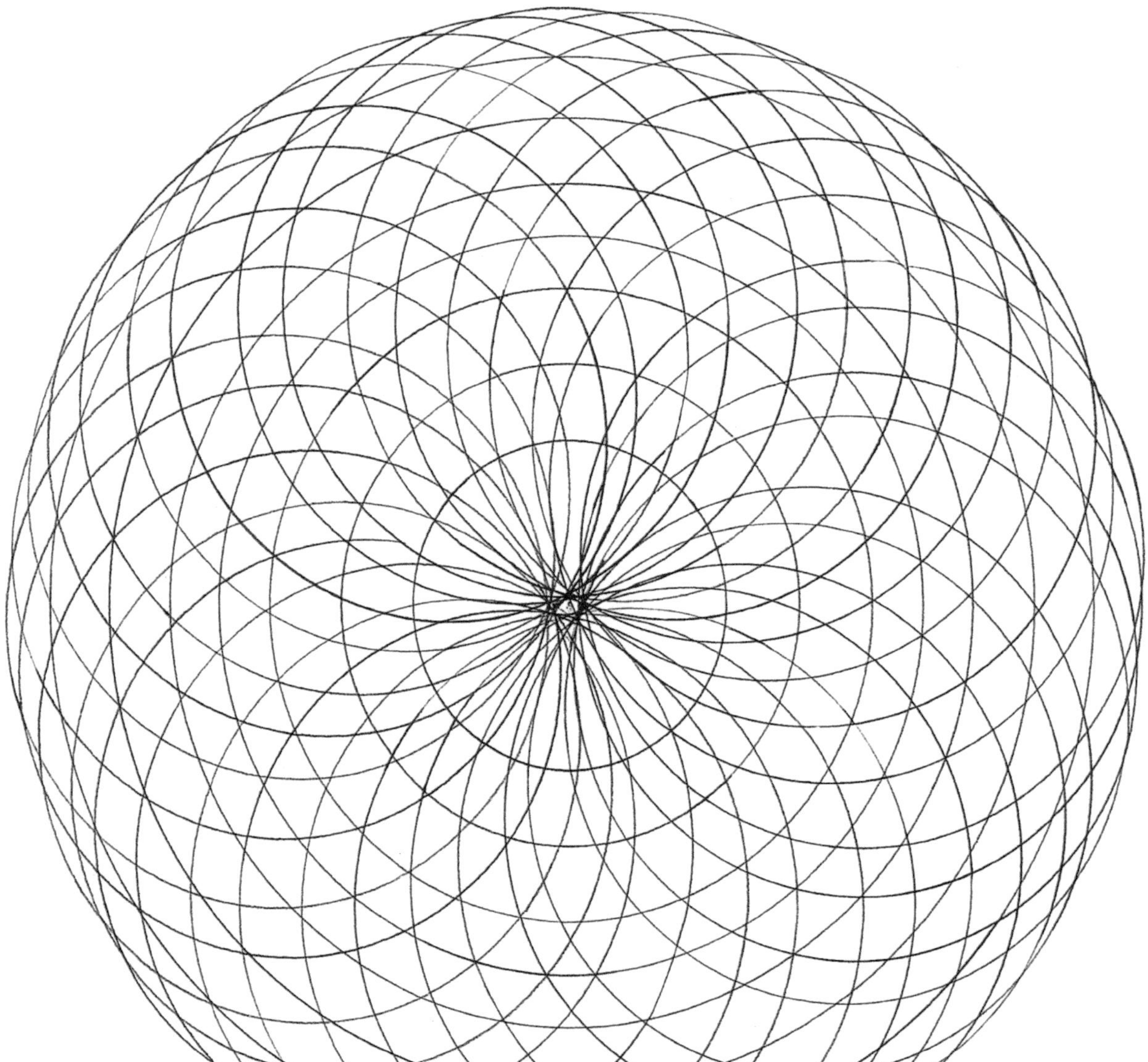

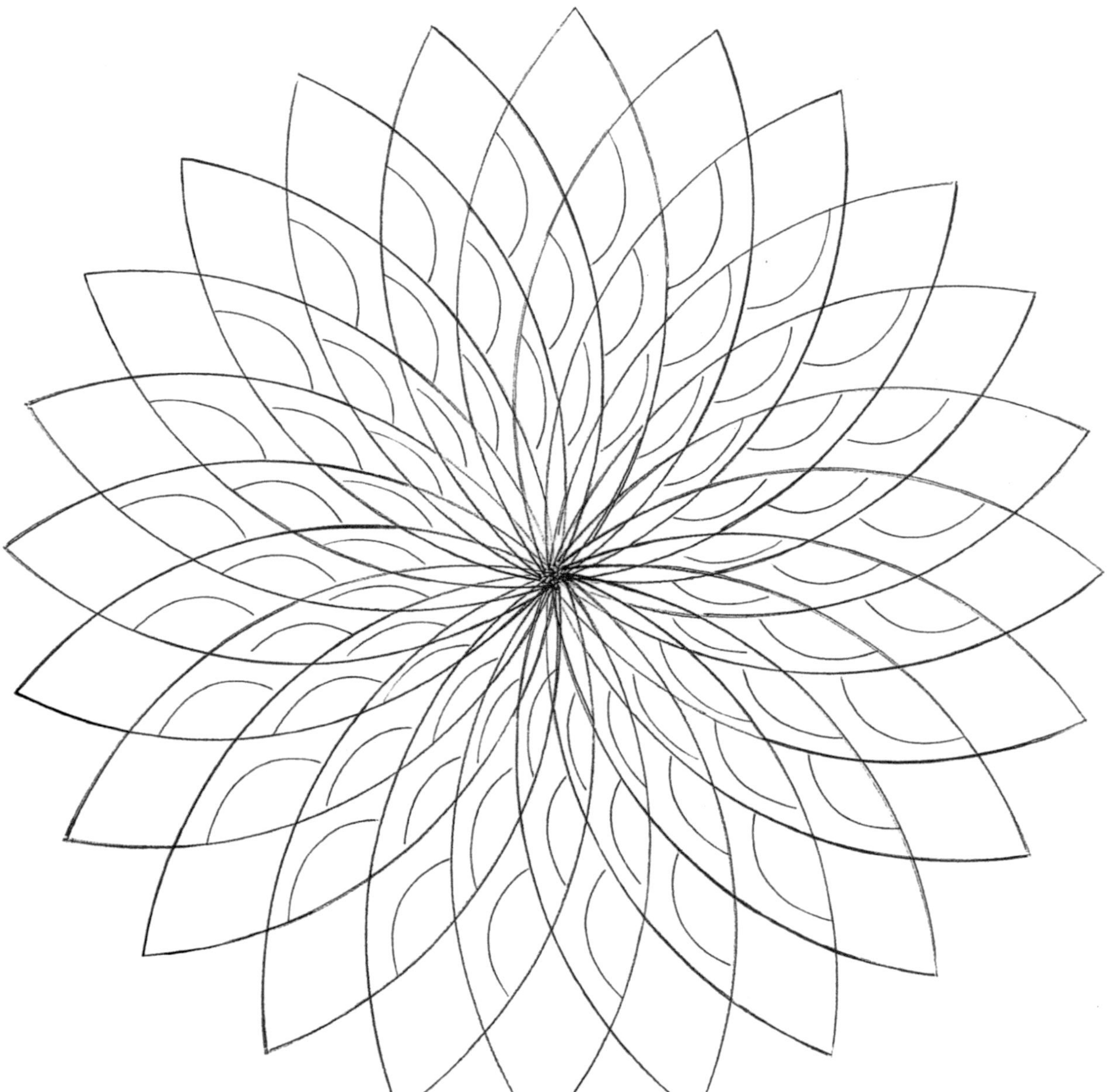

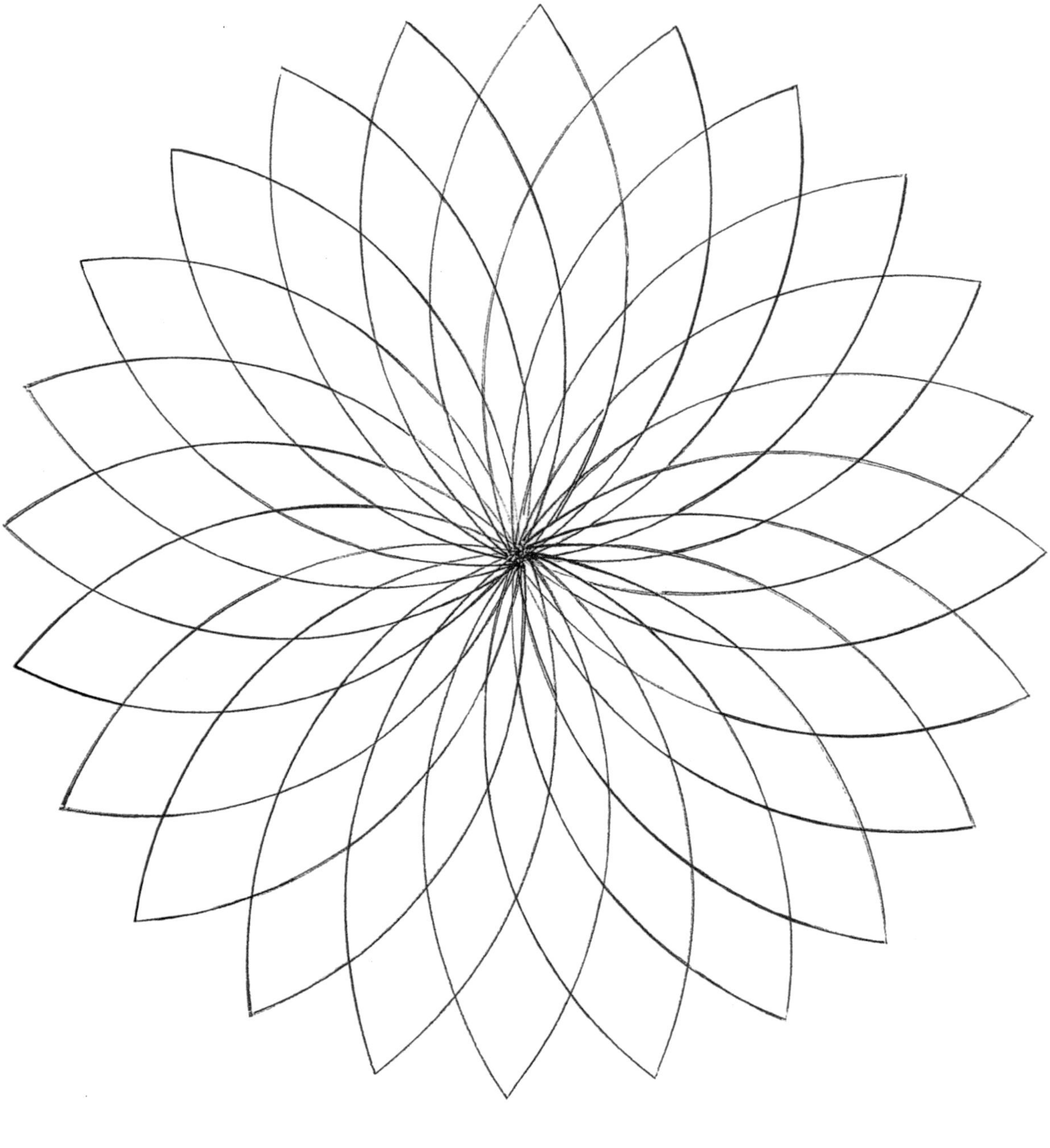

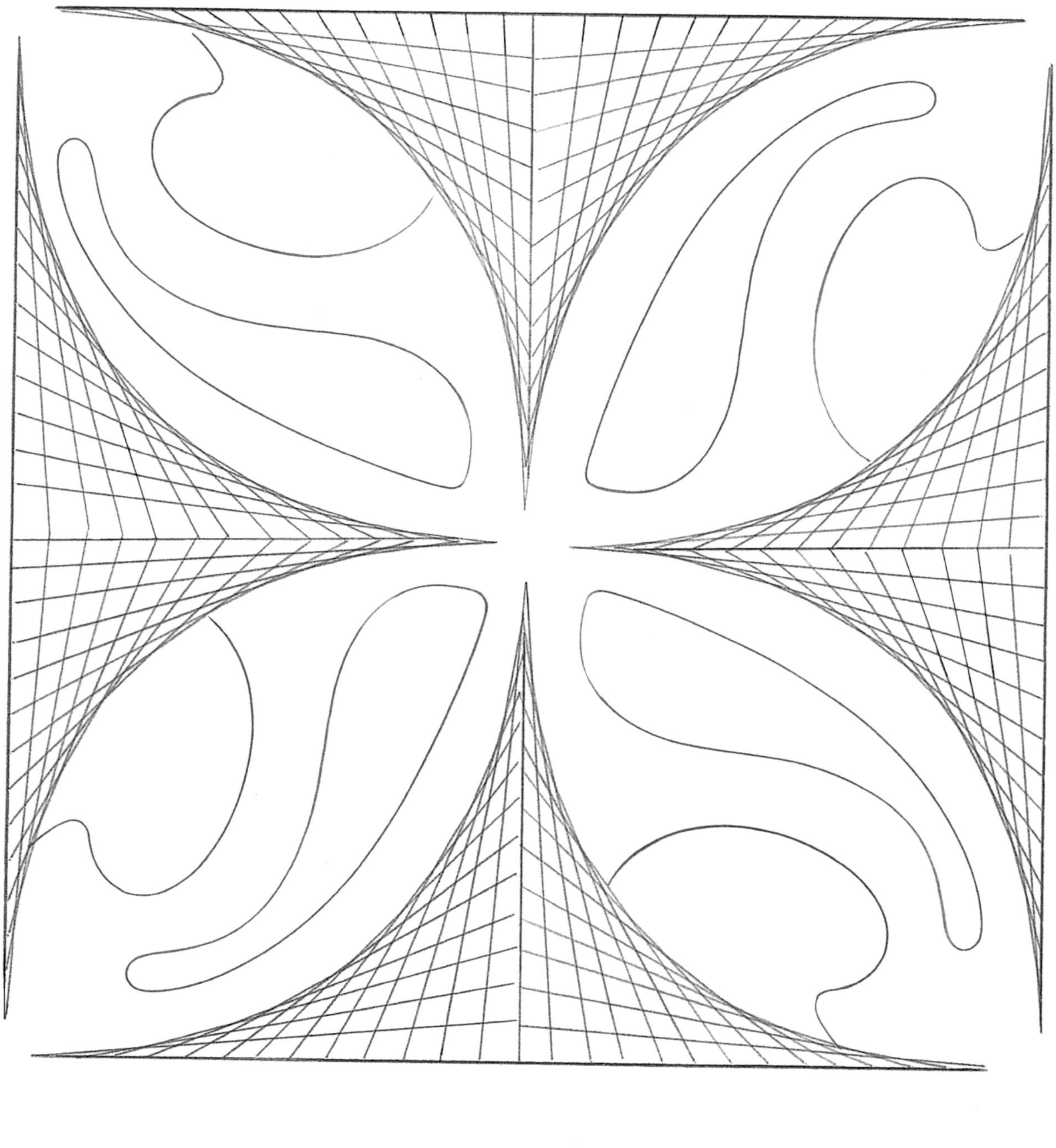

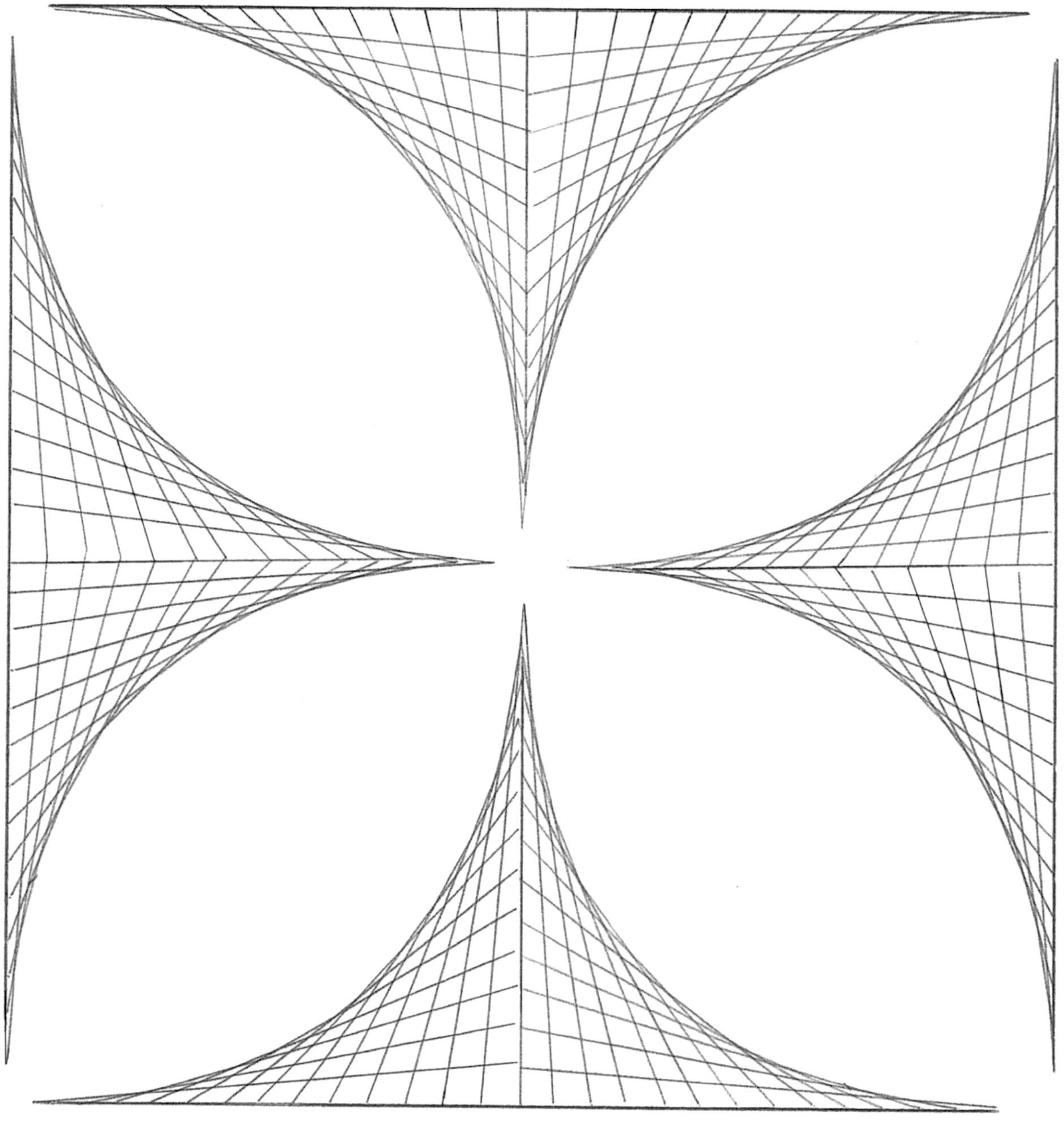

Made in the USA
Monee, IL
05 November 2026

67382335R00040